FORESTS OF THE DINOSAURS

This book is dedicated to my wife Marie, alias the Tree Widow, on account of her husband's many absences and derelictions of duty while engaged in important palaeontological activities, or 'playing with his fossils', the most notable of which was a fossil tree.

FORESTS OF THE DINOSAURS

Wiltshire's Jurassic Finale

JOHN E. NEEDHAM

First published in the United Kingdom in 2011
by The Hobnob Press, PO Box 1838, East Knoyle, Salisbury, SP3 6FA
www.hobnobpress.co.uk

© John E. Needham, 2011

The Author hereby asserts his moral rights to be identified as the Author of the Work.

All rights reserved. No part of this publication may be reproduced, stored in a retrieval system, or transmitted in any form or by any means, electronic, mechanical, photocopying, recording or otherwise, without the prior permission of the publisher and copyright holder.

British Library Cataloguing in Publication Data
A catalogue record for this book is available from the British Library

ISBN 978-1-906978-01-3
Typeset in Minion Pro 12/16pt. Typesetting and origination by John Chandler
Printed by Lightning Source

Cover: A reconstruction by Isabelle Needham of the south-west Wiltshire landscape towards the end of the Jurassic period, based upon the rocks and fossils of the Purbeck Limestone Group of Wiltshire and Dorset.

Contents

List of Tables and Illustrations   7

   Introduction and Acknowledgements   11

1   The Purbeck Fossil Forest of Dorset and Wiltshire   29

2   Wiltshire's Plant and Reptile Bed   67

3   The Dorset Connection   91

4   From Wiltshire to the Americas   123

5   Secrets of Wiltshire's Fossil Forest   155

   Postscript   195

References   203

Index   211

List of Tables and Illustrations

Table 1: Timescale of the last 251 million years. 15
Table 2: Basic classification of Mesozoic vascular plants. 18
Table 3: Basic classification of Jurassic higher vertebrates. 25
Table 4: Past and present Portland Stone/Purbeck stratigraphy. 31
Table 5: Portland and Basal Purbeck succession in the Chilmark and Tisbury area. 71
Table 6: Distribution table of selected Wiltshire Tithonian fossils. 191

Assorted fossil seeds, cone scales, shoots, twigs and wood from the plant and reptile bed, south-west Wiltshire. 13
A cycadophyte trunk from the Fossil Forest horizon of south-west Wiltshire. 20
A monkey-puzzle tree *Araucaria araucana* growing in a garden in south-west Wiltshire. 23
Rocks of the Portland Stone Formation exposed in a disused quarry in the Vale of Wardour. 27
A fossil tree trunk from the Purbeck Fossil Forest of the Isle of Portland, on display at Dorset County Museum, Dorchester. 32
Fossil wood from a temporary exposure of the basal Purbeck Limestone Group beds in the Nadder valley, Wiltshire. 35
Seeds or cone scales from Wiltshire's plant and reptile bed attributable to the species *Carpolithes westi* and a monkey puzzle tree cone scale. 38
The broken end of a 2.5cm diameter piece of fossil wood from south-west Wiltshire's Fossil Forest. 41
A fallen monopodial mountain pine *Pinus mugo* decays on the edge of a forest in the Eastern Pyrenees. 44
Foliage of the Leyland Cypress of the Family Cupressaceae. 47
Horsetails growing along a tributary of the River Nadder, Wiltshire. 51
A fossil fish from the base of the Purbeck Limestone Group, south-west

Wiltshire.	55
A broken sauropod tooth crown cf. *Pelorosaurus* from the Portland Stone Formation of the Vale of Wardour.	59
Small theropod teeth from Wiltshire's plant and reptile bed curve back in a similar manner to *Nuthetes destructor* teeth from Dorset.	62
Turtle, crocodile and dinosaur bones from the Purbeck Limestone Group, Durlston Bay, Dorset.	65
An ammonite from the Wockley Member of south-west Wiltshire's Portland Stone Formation.	70
An assortment of reptile teeth and bones from the plant and reptile bed.	74
An exceptionally well preserved shoot, probably of an extinct conifer, embedded in silt of the plant and reptile bed.	78
Two unusual lengths of silicified wood preserved in the base of the plant and reptile bed, with evidence of the growth pattern.	81
Small crocodile teeth from the Purbeck Limestone Group of Durlston Bay and the plant and reptile bed.	84
Delicate plant shoots from the plant and reptile bed removed in several pieces and reassembled using PVA glue.	87
A pterosaur bone from the Tisbury Member of the Portland Stone Formation.	90
Conifer wood and reptile bone petrifactions from the Portland Stone Formation of south-west Wiltshire.	92
The study of fossil seeds and fruits, such as these specimens from Wiltshire's plant and reptile bed, is known as palaeocarpology.	95
A fossil conifer wood specimen from the plant and reptile bed invites comparison with a specimen from a present-day tree.	98
Silicified specimens of the seed *Carpolithes rubeola* from the plant and reptile bed.	101
Specimens of the remarkable seed species *Carpolithes glans* from the plant and reptile bed.	104
A rare seed of the species *Carpolithes rhabdotus* from the plant and reptile bed.	106
Fruits from the plant and reptile bed enclosing seeds of the species *Carpolithes rubeola*.	108
Probable seed capsules from the plant and reptile bed, each of which would have enclosed two seeds.	111
A carapace fragment of an indeterminate turtle of the extinct freshwater family Solemydidae, from the plant and reptile bed.	114
Teeth of the extinct crocodile *Goniopholis* from the plant and reptile bed and the Cherty Freshwater Member of the Purbeck Limestone Group.	117
The author's daughter Nicole Needham observes the uncovering of a 55cm long piece of silicified wood.	120
Silicified plant fossils from the Henry Mountains of Utah similar to specimens named *Behuninia joannei* by Chandler.	124

✻ List of Tables and Illustrations ✻

Specimens of *Steinerocaulis radiatus*, formerly *Carpolithus radiatus*, from the plant and reptile bed and the Henry Mountains of Utah.	127
Long shoot-short shoot organisation in the larch genus *Larix* and in fossils from the plant and reptile bed.	130
Plant fossils from the Henry Mountains of Utah attributable to *Behuninia provoensis*.	133
Behuninia provoensis-type and possible *Behuninia joannei* short shoots from the plant and reptile bed.	136
Unusual plant specimens from the plant and reptile bed.	138
Cycadophyte trunks such as this probable Wiltshire bennettite trunk are not readily attributable to species level.	141
Short shoots from the plant and reptile bed with *Behuninia provoensis*-type attachments.	144
A small bone, possibly of a frog, from the plant and reptile bed.	147
Camarasaurus and *Camarasaurus*-like teeth from Utah and the plant and reptile bed.	150
Allosaurus and allosauroid teeth from Utah, the gastropod micrite and the plant and reptile bed.	153
Stromatolite mounds on an eroded limestone platform adjacent to south-west Wiltshire's Fossil Forest horizon.	156
Truncated branch bases on a dead mountain pine and Wiltshire's 146 million-year-old fossil tree.	158
Wiltshire's fossil tree as reassembled in November 2008.	161
Details of branch I of Wiltshire's fossil tree and young shoots from the plant and reptile bed.	163
Wiltshire's fossil tree as reconstructed in November 2008, with the background blacked out.	166
Ridges running around the stump of Wiltshire's fossil tree and similar ridges on a fallen mountain pine.	168
A *Behuninia*-like bud or shoot on a conifer branch excavated near the fossil tree.	170
A section of wood from Wiltshire's Fossil Forest with insect borings.	173
Remarkable surface details exposed on roots from Wiltshire's Fossil Forest.	176
A thin fragment from the largest cycadophyte collected by the author exposes internal detail.	178
A highly compressed cycadophyte, surface collected from the Wiltshire Fossil Forest.	181
A group of four cycadophytes on exposed palaeosol of the Wiltshire Fossil Forest.	183
The remains of a young conifer about eight years old, from the Wiltshire Fossil Forest.	186
Fossils sieved from palaeosol collected around the young Fossil Forest conifer illustrated previously.	188

Drawings of lignotubers and/or short shoots from the Tithonian of south-west Wiltshire. 189

One of two rare small silicified cones found in the palaeosol of Wiltshire's Fossil Forest. 192

Map of a 20sq.m area of excavated palaeosol from Wiltshire's Fossil Forest, set out in metre squares and contoured. 196

An artist's reconstruction of the south-west Wiltshire environment during late Tithonian times. 199

Introduction and Acknowledgements

'Countless ages before Man was created, I visited these regions of the Earth and beheld a beautiful country of vast extent. Groves of palms and ferns and forests of coniferous trees clothed its surface, and I saw monsters of the reptile tribe so huge that nothing among the existing races can compare with them.'

Dr Gideon Mantell (1790–1852)

A FEW YEARS AGO the above quote featured near the end of a dramatised documentary on the lives and works of some early British geologists, who were active during the first half of the nineteenth century and referred to as the 'dinosaur hunters'.[1] The words convey as well as any the sense of wonder at the discovery of a previously unknown and unimagined world, a world all the more strange when set against preconceptions, both religious and otherwise, predominant at that time. For men such as Mantell, discoverer of the great herbivorous dinosaur *Iguanodon*, the collection and scientific interpretation of fossils placed them in the spotlight of a changing world view, as yet pre-Darwinian but already introduced to the vast concept of geological time by the works of a pioneering Scottish

---

1   *The Dinosaur Hunters* (2001), Granada Productions for Channel 4.

geologist named James Hutton. In 1797, in *Theory of the Earth*, Hutton had challenged the accepted wisdom of biblical scholars under which the Earth was created in 4004 BC. Instead, erring somewhat in the opposite direction, he saw a world of endlessly repeating geological processes with 'no vestige of a beginning, – no prospect of an end'.[2]

Since those days a diverse array of dinosaur species has been discovered. The reconstruction of some of these in action, through computer generated images based upon the interpretation of skeletons, footprints and coprolites or fossil dung, has helped consolidate their position in the popular imagination. Extraordinary new discoveries are still being made – it is not that many years since feathered bird dinosaurs were discovered in China, and as recently as February 2009 researcher Dr Steve Sweetman from Portsmouth University reported the discovery of forty-eight new fossil species, including 'at least eight new dinosaurs', from the Isle of Wight.[3] With the last decade seeing dinosaur finds running at '30 or so new species ... each year',[4] there seems little doubt that the majority of dinosaur species that have walked this planet remain undiscovered.

As challenging, if not more challenging, than reconstruction of the dinosaurs is the detailed reconstruction of the plant environments in which they lived. The discovery of a dinosaur skeleton is a far more likely event than the discovery of a tree in which more than a stump and some of the trunk has been preserved. Excellent specimens of twigs, cones, seeds and leaves can be found, but seldom attached to the parent plants from which they came. Whole plant reconstruction is thus one of the main challenges of palaeobotany, or the study of fossil plants. As stated in the introduction to a review of British fossil plant conservation sites, 'Most extinct plants have not been reconstructed,

[2] Bartholomew and Morris (1991), 293.
[3] Portsmouth University http://www.port.ac.uk/aboutus/newsandevents/frontpagenews/title,91933,en.html (visited 2 April 2010)
[4] Barrett (2010), 59.

* *Introduction and Acknowledgements* *

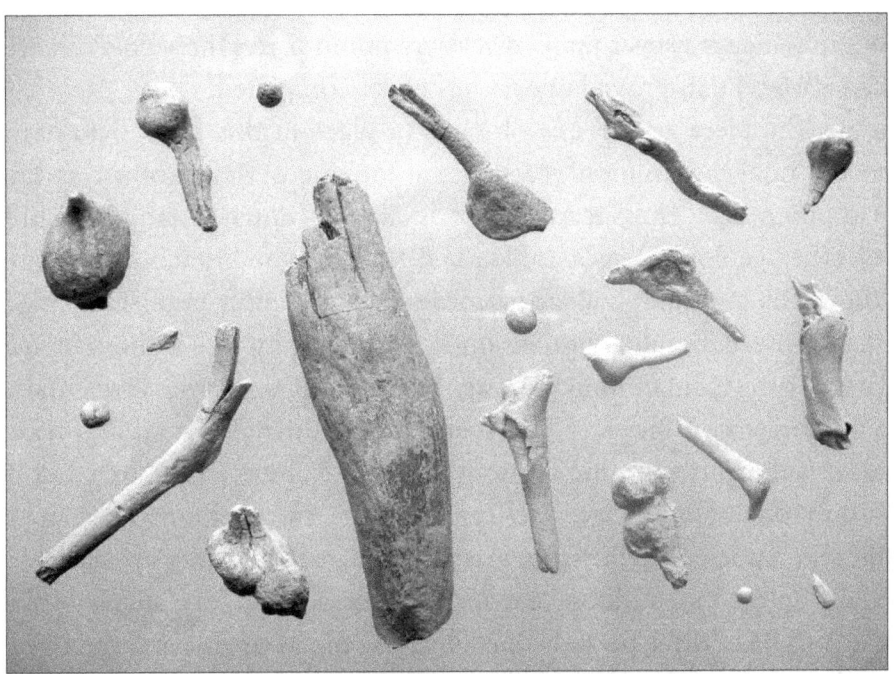

Assorted fossil seeds, cone scales, shoots, twigs and wood. The contents of a plant debris bed, such as these specimens from south-west Wiltshire's plant and reptile bed, demonstrate the problems encountered in whole plant reconstructions. To further complicate matters, leaves are seldom found in association with three-dimensionally preserved plant remains. Specimens: Author's collection. Photo: Author.

because various organs are preserved as fragments that are difficult to link together.'[5]

For this and for a variety of other reasons the preservation of animal and plant remains to an equal standard seldom occurs at the same fossil site, and the complex task of reconstructing terrestrial environments thus often depends on integrating information from various localities. The discovery of a fossil that links one site to another, sometimes across different continents, is a highly satisfying experience. It is like working on a jigsaw puzzle on a grand scale, in which years of searching occasionally allow one more piece to be placed in position,

5   Cleal and Thomas (2001), 10.

but in which however many pieces are put in place, the whole picture can always be improved upon - but never completed.

The piece or pieces to be put in place in this book date back to the final few million years of the Jurassic period, known as the Tithonian age. The succession of rocks laid down during this and all other geological ages are studied through the branch of geology known as stratigraphy, and according to the latest published chart of the International Commission on Stratigraphy[6] the Tithonian age lasted from 150.8 million years ago (mya) until 145.5mya, give or take 4 million years either way. This span of approximately 5.3 million years may well dwarf the time of modern man's existence on Earth, but it is only one of three ages that make up the 15.7 million year Upper Jurassic epoch, which in turn is only one of the three epochs that make up the 54.1 million year Jurassic period. The Jurassic is in turn the middle of three periods that comprise the Mesozoic era, the 185.5 million year 'Age of Reptiles' ('age' being used here in a colloquial rather than geological sense) that began with the Triassic period and drew to a sudden close with the extinction of the dinosaurs at the end of the Cretaceous period. That era was followed by the Cainozoic era, the 65.5 million year (so far) 'Age of Mammals'. Some of the terms used, it should be noted, have changed over the decades, and these days geological literature can fast become out of date. Early geologists such as Mantell spoke of 'Secondary strata' rather than of 'Mesozoic strata', and the term 'Tertiary', covering all but the last approximately 2½ million years of the Cainozoic era, has been 'replaced in current global chronostratigraphical parlance'[7] – i.e. it is technically redundant, although in practice still widely used. To minimise possible confusion where references to earlier works are made, some archaic terms are included in Table 1.

---

6   Based on Gradstein, Ogg and Smith (eds) (2004).
7   Hooker, Cook and Benton (2005), 69.

* *Introduction and Acknowledgements* *

Standard (faunal) time divisions				Floral time divisions	Time-scale	Archaic terms
Cainozoic Era (Age of Mammals)	Neogene Period			Cainophytic (Age of Flowers)	0-2.588mya	Quaternary*
	Palaeogene Period				2.588-65.5mya	Tertiary
Mesozoic Era (Age of Reptiles)	Cretaceous Period	Upper		Mesophytic (Age of Conifers)	65.5-99.6mya	Secon-dary
		Lower			99.6-145.5mya	
	Jurassic Period	Upper			145.5-161.2mya	
		Middle			161.2-175.6mya	
		Lower			175.6-199.6mya	
	Triassic Period	Upper			199.6-251.0mya	
		Middle				
		Lower				

*Retention proposed as sub-era.

Table 1: *Time scale of the last 251 million years. Estimated margins of error on Mesozoic dates range from ±0.3 million years to ±4.0 million years.*[8]

The standard divisions of geological time are based upon changes in the animal life revealed in the fossil record, but divisions can also be made based upon changes in plant life, and the two do not necessarily correspond. The middle period for plant life on Earth, the Mesophytic or 'Age of Conifers', begins at the same time as the Mesozoic era with a great extinction event around 251mya, but it draws to a close around 150 million years later in the middle of the Cretaceous period with the rapid radiation of the angiosperms or flowering plants and the arrival of the Cainophytic or 'Age of Flowers'. The great dinosaurs of the Jurassic and the Lower Cretaceous such as *Diplodocus*, *Iguanodon* and *Megalosaurus* thus roamed through a landscape as yet devoid of true flowers, but when *Tyrannosaurus rex* ruled – or, as now seems more likely to have been the case, scavenged – in the Upper Cretaceous the flowering plants had seized

8 Standard divisions and time scale after Gradstein, Ogg and Smith (eds) (2004), Mesophytic from Kenrick and Davis (2004), 141.

the dominant position in the plant kingdom. Rather than share the fate of the dinosaurs around 65.5mya, when the Age of Reptiles drew to a rapid conclusion with their extinction, the flowering plants successfully continued their remarkable diversification to what is now 'at least 240,000 extant species'.[9]

At this point it is worth taking a brief look at the position of the conifers and flowering plants, along with some other significant plant groups within the plant kingdom. The kingdom comprises a number of divisions or phyla, subdivided into classes, orders and families, although when it comes to the details there is more than one proposed classification available – in fact there are lots of them. A recently published marathon work on palaeobotany notes that 'naming plants still represents a highly subjective exercise', and of its own classification scheme it adds: 'Some readers may wish to adapt the plant groups ... to a system with which they feel more comfortable.'[10] The scheme followed here for the most part is that adopted by the UK-published book *Mesozoic and Tertiary Palaeobotany of Great Britain*[11], although even within this one work there are inconsistencies between the contributory authors. Under the classification five divisions of vascular land plant living in the Mesozoic era are recognised. Vascular plants are those with conductive tissues for the transport of water and nutrients. Two of these, the Lycophyta and the Sphenophyta, had seen better days during the latter part of the preceding Palaeozoic era, or in plant terms the Palaeophytic. Around 300mya tree-sized representatives of both groups were abundant in the swamps of the Carboniferous period, from which we have inherited much of the world's coal deposits. They have survived to today in greatly reduced circumstances, both in terms of size and diversity. The Lycophyta includes the clubmosses and quillworts, of which the fir clubmoss

---

9   Cleal and Thomas (2001), 5.
10 Taylor, Taylor and Krings (2009), 41-2.
11 Cleal and Thomas (2001), 10-12.

found today in parts of the English Pennines is an example. The Sphenophyta includes the horsetails, common in many marshy riverside locations in England. The Pteridophyta is the division representing the ferns, which retain a reduced but significant role in today's vegetation following the rise of the angiosperms.

The next division is the Gymnospermophyta. Gymnosperms are plants with open as opposed to enclosed seeds, typically but not always produced in cones, and include several large-leafed Mesozoic orders which are mostly extinct. Five such orders, all of significance within UK Jurassic strata, are looked at within the classification scheme employed in the US published book *Paleobotany: The Biology and Evolution of Fossil Plants*.[12] This is the only major diversion from the scheme employed in *Mesozoic and Tertiary Palaeobotany of Great Britain*, and is made in the interests of clarity. Two of these, Orders Caytoniales and Corystospermales, are attributed to an informal group incorporating plants of uncertain inter-relationships under the heading 'Mesozoic seed ferns'[13] or Pteridospermophyta (no relation to Pteridophyta or true ferns). Named after Cayton Bay in Yorkshire, the caytonias had leaves with radiating leaflets and bore 'fleshy and edible fruits', but there is 'no reliable reconstruction of the whole plants although they were most probably shrubs'.[14] The less abundant corystosperms are represented by foliage species with 'leathery fern-like leaves'.[15] Order Czekanowskiales, grouped under 'Other gymnosperms'[16] or 'Gymnosperms with obscure affinities', is represented by plants with 'persistent leaves born on ... deciduous spur shoots' that produced seeds in bivalve capsules.[17] The two remaining

---

12   Taylor, Taylor and Krings (2009), 1028.
13   Taylor, Taylor and Krings (2009), 621.
14   Thomas and Batten (2001a), 45.
15   Thomas and Batten (2001a), 47.
16   Taylor, Taylor and Krings (2009), 1028.
17   Taylor, Taylor and Krings (2009), 765–7.

orders, Order Cycadales and Order Bennettitales, are the 'cycads' or cycadophytes (Cycadophyta), of which 'true' cycads are restricted to the former.

Divisions of the Plant Kingdom	Classes (or Phyla)* to which reference is made within this book	Orders to which reference is made within this book	Existing representatives
Lycophyta	Lycopsida		Clubmosses, quillworts
Sphenophyta	Equisetopsida	Equisetales	Horsetails
Pteridophyta	Filicopsida		Ferns
Gymnospermo-phyta	(Pteridospermophyta)	Caytoniales	
		Corystospermales	
	(Cycadophyta)	Cycadales	'True' cycads, e.g. sago palm
		Bennettitales (or Cycadeoidales)	
	(Other Gymnosperms)	Czekanowskiales	
		Gnetales	Joint firs, mormon teas
	Pinopsida	Pinales (or Coniferales)	Monkey puzzle trees, pines, yews, redwoods, cedars, junipers
		Ginkgoales	Maidenhair tree
Angiospermae			Flowering plants

*Phyla in Taylor, Taylor and Krings (2009) are often approximate equivalents to Classes in Cleal and Thomas (2001).

Table 2: Basic classification of Mesozoic vascular plants[18]

The palm-like 'true' cycads survive to this day, and a living example is the sago palm of southern Japan. Further examples of

---

18   Based on Cleal and Thomas (2001) and Taylor, Taylor and Krings (2009).

※ *Introduction and Acknowledgements* ※

these tropical plants are found in many other parts of the world from Central America to South Africa and Australia. In contrast the cycad-like plants of Order Bennettitales, also known commonly as Order Cycadeoidales, became extinct around the end of the Mesozoic era, possibly because they possessed reproductive structures in which the male pollen organs were fully enclosed. Boring insects such as beetle larvae could have been secondary pollinators, but an otherwise self-pollinating system 'would have contributed to a ... population that would not be well-adapted to environmental changes'.[19] Unencumbered by such problems, the 'true' cycads were able to survive the spread of the angiosperms.

Differences in reproductive structures highlight what has been described as the possibly 'unnatural status of the old group "cycadophytes"',[20] and the bennettites have been seen as more closely related to woody shrubs of the extant Order Gnetales, such as joint-firs or mormon teas, than to the 'true' cycads.[21] Despite this, the similarities between the long frond-like leaves with numerous leaflets or pinnae, only distinguishable under the microscope, and between the diamond-shaped leaf bases on the trunks and branches, make the Cycadophyta a convenient informal grouping. The widespread distribution of cycadophyte fossilised leaves and to a lesser extent trunks in Jurassic plant-bearing strata and their distinctive appearance has led to the Jurassic period being 'referred to as the Age of Cycads'.[22] These fossils also range well beyond the Jurassic period, and it has thus even been said that 'the Mesozoic has come to be known as the "age of the cycadophytes"'.[23] Such is the impression they can make that the American palaeobotanist Lester F. Ward, working during the

---

19 Taylor, Taylor and Krings (2009), 732.
20 Stewart and Rothwell (1993), 363.
21 Cleal and Thomas (2001), 11.
22 Daniels and Dayvault (2006), 98.
23 Stewart and Rothwell (1993), 346.

A cycadophyte trunk from the Fossil Forest horizon of south-west Wiltshire seen from the underside. A helix of diamond-shaped leaf bases radiates from the central growing point. With a diameter of 23cm, the specimen has probably undergone considerable compression before fossilisation. Specimen: Author's collection. Photo: Steve Clifford.

latter part of the nineteenth and beginning of the twentieth century, wrote: 'Cycads are to the vegetable kingdom what Dinosaurs are to the animal, each representing the culmination in Mesozoic times of the ruling Dynasties in the life of their age.'[24] Their fossils also made a marked impression on Mantell, who misinterpreted them as 'groves of palms' from a time when true palms had yet to evolve.

The remaining gymnosperms are grouped into two orders, the Pinales or Coniferales and the Ginkgoales. The latter was globally widespread during the Mesozoic era but is now represented by one solitary Far Eastern species, the maidenhair tree *Ginkgo biloba*, whose

---

24 Cited Taylor, Taylor and Krings (2009), 703.

anti-oxidant rich leaf extract reputedly improves circulation to the brain and other vital organs. The former consists of the conifers, and eight conifer families are recognised under the classification system followed here. All but one of these have present-day representatives and all but one were to be found towards the end of the Jurassic period. The Podocarpaceae is represented today by the podocarps of the southern hemisphere, including the giant kauri tree of New Zealand, but occurred in both hemispheres in Jurassic times. The Araucariaceae, most famously represented by the monkey-puzzle tree, now has a limited southern hemisphere distribution but was globally widespread in Jurassic times. The remarkable Wollemi pine *Wollemia nobilis*, discovered in Australia in 1994 and heralded as a living fossil from the age of the dinosaurs, belongs to this family. The Taxaceae is represented by the yews and has in the past been classified as a separate order, Order Taxales, because clearly defined seed cones are absent and the trees 'bear their seeds terminally on short lateral shoots'.[25] The Taxodiaceae is represented by the redwoods, the Cephalotaxaceae is a small family now living in Asia that includes the Chinese plum-yew, and the Cupressaceae includes the cedars and junipers. The Cupressaceae has been described as 'the only modern (conifer) family that was relatively late to appear ... whose oldest known examples are Tertiary in age'.[26] This would make the family less than 65.5 million years old, although a related source mentions the reported occurrence in the 170 million year old Middle Jurassic strata of Yorkshire of 'the fossil wood genus *Cupressinoxylon*, which probably belongs to this family'.[27] More recent classifications have included in the Cupressaceae 'taxa previously in the family Taxodiaceae',[28] which definitely lived during the Jurassic period. The pines make up the Pinaceae, and the

25   Stewart and Rothwell (1993), 413.
26   Cleal and Thomas (2001), 5.
27   Thomas and Batten (2001a), 54.
28   Taylor, Taylor and Krings (2009), 850.

Cheirolepidiaceae is an extinct family that was widespread during the Mesozoic era. Members of the Cheirolepidiaceae have also been referred to as 'transitional taxodioids',[29] and the foliage 'has historically been included in a number of conifer families, most commonly, the Cupressacae, Taxodiaceae, and Araucariaceae'.[30] Other conifer families extant during the Mesozoic era have been recognised, of which the only one of likely relevance here is the Pararaucariaceae represented by Jurassic fossil cones from South America. The family is an intermediate one, as the cones incorporate features of both the Cupressaceae and Pinaceae.[31]

This just leaves the Angiospermae or angiosperms, distinguished anatomically by a number of features including enclosed seeds. This is the group that has come to dominate the plant kingdom today, providing not only much of the vivid variety of colour to be found in nature but also the grasses and broad-leafed trees. Examples are well known from the Lower Cretaceous, when they were a minority group in the plant world. The flora from the Lower Greensand Group of southern England dates back to this epoch and was described in 1915 by Marie Stopes,[32] better known for her progressive approach to family planning and women's rights but equally adept in the field of palaeobotany. What she described was a Mesophytic flora about 110 million years old, including ferns and bennettites along with twenty-seven conifer and five angiosperm species. The flowering plants were thus already well established many millions of years before the end of the 'Age of Conifers', but earlier evidence of angiosperm traits goes way back into the Jurassic period. Jurassic fossil plant sites, and especially those where previously undescribed material is to be found, thus raise an exciting possibility: could they offer any

---

29 Stewart and Rothwell (1993), 429.
30 Taylor, Taylor and Krings (2009), 833.
31 Taylor, Taylor and Krings (2009), 861.
32 Thomas and Batten (2001b), 152.

* *Introduction and Acknowledgements* *

A monkey-puzzle tree Araucaria araucana growing in a garden in south-west Wiltshire. The tree belongs to the family Araucariaceae, which was widespread during the Mesophytic but now occurs naturally within limited areas of the Southern Hemisphere. The Australian Wollemi pine also belongs to this family.
Photo: Author.

new evidence on angiosperm origins? They hardly ever do, but it is always a tantalising possibility because, in the words of Natural History Museum palaeobotanists Professor Paul Kenrick and Dr Paul Davis, 'The origin of flowering plants is one of the most widely discussed and enduring mysteries in the history of life on Earth.'[33] Various gymnosperm groups such as ginkgos, pteridospermophytes and cycadophytes have been associated with angiosperm evolution, although more as distant cousins than as ancestors. The mystery will more than likely endure for some time to come, as 'there is still no unambiguous fossil that can be used to illustrate a pre-Cretaceous angiosperm.'[34]

Although it is the plants that tend to take centre stage over the following pages, their significance is inevitably enhanced by the extraordinary array of wildlife with which they co-existed – an assembly of creatures that has captured the human imagination since the earliest scientific interpretations of dinosaur remains. In Tithonian times there were amphibian forebears of our frogs and salamanders, and there were reptile ancestors of the turtles, snakes, lizards and crocodiles, but it is the members of five extinct orders of reptile that give such a distinctly alien feel to this long gone world.

The Pterosauria were the flying reptiles, from those the size of small birds to those with wingspans measured in metres. The Saurischia or lizard-hipped dinosaurs included the huge herbivorous sauropods such as *Diplodocus* and the predatory carnivorous theropods such as *Allosaurus*. One group of theropods, the maniraptors, evolved into today's birds. The Ornithischia or bird-hipped dinosaurs included herbivores with plant-cutting beaks such as Mantell's famous *Iguanodon* and armoured or plated forms including nodosaurs, ankylosaurs and stegosaurs. In the sea the

---

33   Kenrick and Davis (2004), 186.
34   Taylor, Taylor and Krings (2009), 877.

ichthyosaurs and plesiosaurs were the sea monsters of their day. Not to be forgotten, and living in the shadow of the dinosaurs, were the many small ancestors and cousins of today's mammals, biding their time through tens of millions of years of dinosaur ascendancy.

Class	Order	Fossil and extant representatives
Amphibia	Anura	Frogs and toads.
	Caudata	Salamanders and newts.
Reptilia	Testudines	Tortoises and turtles.
	Sphenodontia	Extinct sphenodonts and tuatara.
	Squamata	Mosasaurs, snakes and lizards.
	Crocodilia	Extinct 'mesosuchian' crocodiles, crocodiles and alligators.
	Pterosauria	Flying reptiles
	Saurischia	Sauropod herbivorous dinosaurs such as *Diplodocus*. Theropods, including carnivorous dinosaurs such as *Allosaurus*. The theropod sub-group Maniraptora includes by evolutionary descent all birds, traditionally classified in Class Aves.
	Ornithischia	'Bird-hipped' herbivorous dinosaurs such as iguanodonts and the armoured nodosaurs, ankylosaurs and stegosaurs.
	Ichthyosauria	Ichthyosaurs.
	Plesiosauria	Plesiosaurs.
Synapsida	Unranked group Mammaliaformes	Various extinct ancestors and cousins of present-day mammals, and by evolutionary descent all mammals, traditionally placed in Class Mammalia.

Table 3: Basic classification of Jurassic higher vertebrates.

The English county of Wiltshire has its own special role to play in the search for evidence of the world of the dinosaurs at the end of the Jurassic period, and it is the purpose of this book to explore that role through early geological accounts, through comparisons with finds from other continents, through scanty accounts of largely unpublished discoveries made over twenty-five years ago, and through an examination of finds made by the author during the past few years. In an adventure into this world of so many unknowns, the Portland Stone and Purbeck Limestone floras of south Wiltshire will be seen as fundamental in linking the Tithonian environments of Europe with those of the western USA and, to a lesser extent, with those of Argentina. The significant plant fossils are almost exclusively 'petrified' specimens or petrifactions, ranging in size from little more than 1mm across to almost 12m in length. American fossils from this age include the famous dinosaur remains at the Dinosaur National Monument on the Utah/Colorado border, and although similarly aged dinosaur remains from south-central England are far less spectacular, in the reconstruction of these environments everything from the smallest tooth to a complete dinosaur skeleton and from a conifer seed or shoot to an almost complete tree can play a significant role. In interpreting the terrestrial world of the Tithonian age, Wiltshire is thus a major source of evidence, and even Dorset with its famed Jurassic Coast World Heritage Site has to stand in awe of its neighbouring county to the north, previously seen as little more than an also ran.

Wiltshire fossils collected by the author and discussed in this book, which are now either in the Natural History Museum (NHM) in London or remain in his own collection, were found at localities in the Vale of Wardour in the south-west of the county between 2002 and 2010. Access to sites on private land was made possible through the author's work, and permission to collect was kindly given by landowners. Specific localities are not identified. Portland Stone and Purbeck Limestone form the bedrock around several villages along the River Nadder and its side valleys, including Wardour, Tisbury, Fonthill Gifford, Fonthill Bishop, Upper Chicksgrove, Lower Chicksgrove, Chilmark and Teffont Evias. Geological exposures occur within this area at various disused quarry workings, at one working quarry and one recently working mine, and at periodic temporary exposures usually connected with construction work.

Introduction and Acknowledgements

Rocks of the Portland Stone Formation exposed in a disused quarry in the Vale of Wardour. Stone extraction was formerly widespread, with up to forty quarries recorded from the Tisbury area alone. The decline of the industry and quarry infill has greatly reduced the number of geological exposures available. Photo: Author.

Finally, before the reader passes through the gateway into Wiltshire's lost Jurassic world, a few acknowledgements are in order to those whose help has been invaluable. They include property owners who gave permission to collect; Professor Paul Kenrick at the NHM who offered advice and assistance and arranged for sectioning of specimens; all those at the NHM who processed and identified specimens illustrated over the following pages, including Dr Paul Barrett, Dr Angela Milner, Sandra Chapman and Dr Martin Munt; John Chandler of Hobnob Press who made this entire project possible and designed the layout; Simon Fletcher who as editor for Hobnob Press gave advice and assistance and patiently extended the agreed deadline for this book; and Johnny and Jenny Williamson and Elizabeth Birch who read through and commented on the text. Geologist Richard Dayvault of Colorado always offered his expert opinion on specimens based upon his knowledge of plant fossils from the western USA, drawing also on the extensive expertise of Professor William D. Tidwell of Brigham Young University, and Dr Roy Clements provided a detailed report on molluscs he kindly agreed to identify. The pterosaur bone illustrated in Chapter Two was given to the author by its finder, Nick Porter, and the dinosaur centrum illustrated in Chapter One was given to the author by John Russell. Mr and Mrs J. Moore gave permission for their monkey-puzzle tree to be illustrated, and Dorset County Museum gave permission for publication of the photograph of their Purbeck tree. For help in finding, storing and loading the fossil tree described in Chapter Five thanks are due to the author's daughter Isabelle Needham, Brian Annetts, Humphrey and Solveig Stone, Gary Domoney and Chris Palmer at Domoney Woodwork, and others who wish to remain anonymous. The forest reconstruction on the front of this book and the final illustration were designed and drawn by Isabelle Needham, Steve Clifford took and edited several of the photographs, and the NHM allowed joint copyright on their photographs of cones donated to them by the author. Last but not least, for teaching many of the skills needed to approach the task of researching and collating a work such as this, the author would like to thank the Open University.

1
THE PURBECK FOSSIL FOREST OF DORSET AND WILTSHIRE

IN DORSET, above the massive limestone beds of the Portland Stone Formation, extensively quarried over the centuries for building stone on the Isle of Portland and elsewhere, there lies a complex succession of limestones, clays and marls. Professor William Buckland of Oxford, pioneering geologist and Anglican cleric, was the first to describe this succession in 1818 as the Purbeck Beds, after the Isle of Purbeck in Dorset where some of the most significant outcrops are to be found. Latest literature refers to the Purbeck Beds as the Purbeck Limestone Group, and this title now appears to be the scientifically accepted one. To confuse matters, Purbeck Group, Purbeck Formation and Purbeck Limestone Formation have also been used in recent decades.[35]

The Portland Stone Formation and the Purbeck Limestone Group outcrop not only along the Dorset coast, but also in Wiltshire in the Vale of Wardour and in the Swindon area. The succession from the one formation to the other marks a major transition from a marine to a predominantly non-marine environment, as the sea that had covered a large area of south-central England since pre-Jurassic times

---

35 Ensom (2002), 7.

receded. The marine environment during deposition of the Portland Stone was one of calm open sea in Dorset, but in Wiltshire there was considerable current activity and evidence of a nearby coastline. The environments in which the Purbeck Limestone Group was laid down included episodic intertidal marine and intertidal brackish water zones[36] and freshwater to brackish lagoons where at times seasonal evaporation led to hypersaline conditions. Dry mudflats emerged periodically and longer sub-aerial exposure led to the formation of soil beds.

The entire Portland Stone Formation was laid down during the Tithonian age and so, by what seems to be general agreement, were the basal beds of the Purbeck Limestone Group. It is within these latter beds that the remains of *in situ* fossilised trees known collectively as the Purbeck Fossil Forest are to be found. These basal beds belonged to what was known quite happily for decades as the Lower Purbeck Beds, although in Dorset at least they are now part of the Lulworth Formation. Purbeck stratigraphy is a complex matter and Table 4 below outlines the basic units as recognised before 1975 and after 1996. The point of the boundary between Tithonian age and Berriasian age deposits, and hence between the Jurassic and the Cretaceous, has been debated since the nineteenth century, but whatever the exact answer it is fairly safe to say that the Purbeck Fossil Forest formed during the latter part of the Tithonian.

At a more specific level the Purbeck Fossil Forest is found in a bed known in Dorset as the Great Dirt Bed, which was identified as a fossil soil or palaeosol in 1826. Palaeosol sounds better than dirt, but the latter is a colloquial Dorset synonym for soil, and it was common practice in the nineteenth century for geologists to adopt quarrymen's terms to describe specific strata. Fossil plant remains also occur in an underlying bed known as the Lower Dirt Bed. In

---

36   Norman and Barrett (2002), 161.

1828, when palaeobotany was in its infancy – according to the *Code of Botanical Nomenclature*, 'palaeobotanical literature commenced in 1820'[37] – Buckland wrote a paper on bennettites from the dirt beds of the quarries on the Isle of Portland. Two species were named, *Cycadeoidea microphylla* and *Cycadeoidea megalophylla*.

Stratigraphy (pre-1975)		Stratigraphy (post-1996)*		Age (post-2004)
Purbeck Beds	Upper Purbeck Beds	Purbeck Limestone Group	Durlston Formation (Dorset)	Berriasian Age (Lower Cretaceous)
	Middle Purbeck Beds		Lulworth Formation (Dorset) Fossil Forest in basal beds	Age uncertain: Berriasian or Tithonian
	Lower Purbeck Beds, Fossil Forest in basal beds			Tithonian Age (Upper Jurassic)
Portland Stone or Upper Portland Beds		Portland Stone Formation		

*Based on papers by W.G. Townson (1975), R.G. Clements (1993) and R.K. Westhead and A.E. Mather (1996).

Table 4: Past and present Portland Stone/Purbeck stratigraphy.

The bennettite trunk petrifactions had become petrified through a process known as silicification, as had the stumps and trunks of large coniferous trees preserved in the dirt beds. In *Geology and Mineralogy considered with Reference to Natural Theology* by Buckland, published in 1836, an illustration bears the caption: 'Appearance of trunk and roots of large Coniferous trees, and of trunks of Cycadites, in the black earth, which formed the soil of an ancient Forest in the Isle of Portland.' This work had been commissioned as part of a major

37 Cited Thomas (1991), 7.

undertaking to reconcile science with religion, under the umbrella title of the *Bridgewater Treatises on the Power, Wisdom and Goodness of God as manifested in the Creation* – a fascinating insight into the thinking of a man of science and religion years before Darwin's *Origin of Species* appeared on the scene. Meanwhile, in that same year of 1836, a geologist named William H. Fitton illustrated a 7.2m silicified coniferous stump and trunk from a quarry on the Isle of Portland. Compression of the trunk indicated that the tree had fallen over before becoming fossilised. Bringing the Purbeck Fossil Forest more into the public eye, the Royal Botanic Society in the 1840s placed several large pieces of fossil conifer wood from the Isle of Portland by a lake in London's Regent's Park. Mantell meanwhile joined the ranks of those geologists writing on the subject, and the findings of the early researchers were summarised in 1983 as follows: 'They discovered erect tree stumps in their original growth positions, often over 1m in height ... Numerous fallen trunks and branches were observed on the fossil soil, some over 10m in length and 1.3m in diameter. The erect stumps were shown to have roots which spread laterally through the soil ...'[38] Mantell and others believed the density of these fossils indicated that they were the remains of a true forest.

Early attempts to interpret the Purbeck climate were also made, with Buckland and

The author's daughter Isabelle Needham stands by a fossil tree trunk from the Purbeck Fossil Forest of the Isle of Portland, on display at Dorset County Museum, Dorchester. A similar specimen measuring 7.2m in height was first described by William H. Fitton in 1836. Photo: Author.

38   Francis (1983), 278.

Fitton correctly deducing that the climate was warmer than at present on the incorrect interpretation of the bennettites as 'true cycads', analogous with the extant tropical cycads. Fossils of the latter group are apparently absent from the Purbeck Fossil Forest, although not necessarily because they did not grow there. A list of plants from the Middle Jurassic deposits of the Cleveland Basin of Yorkshire, where leaves are frequent and most fossils are compressions rather than three-dimensional petrifactions, includes thirty-three species from Order Cycadales and fifty-five from Order Bennettitales.[39] Upper Jurassic deposits with leaf impressions 'have yielded higher percentages of modern cycad leaves than extinct cycadeoid leaves, suggesting that the modern cycads were at least as common'.[40] To the author's knowledge determinate leaves of neither group have been recorded from any horizon within the Purbeck Limestone Group, as conditions did not favour such leaf preservation. Although the lack of true cycad trunks in the Purbeck Fossil Forest *could* be because they did not grow there, it could equally be because 'the true cycad trunks did not fossilize as readily as the extinct variety'.[41]

Despite impressive exposures of the Purbeck Fossil Forest on the Isle of Portland and at Lulworth, the material was not easily classifiable into new species other than in the case of the more distinctive bennettites. Further work on these was published in 1870 by William Carruthers in a paper entitled *On fossil cycadean stems from the secondary beds of Britain*, and in 1897 Albert C. Seward described a new species, *Cycadeoidea gigantea*, on the basis of a specimen with a height of 1.18m and a girth of 1.7m. It was to be seventy-eight years before another plant species from the Purbeck Fossil Forest would be described.

Other nineteenth-century plant finds included a conifer

---

39 Thomas and Batten (2001a), 33–5.
40 Daniels and Dayvault (2006), 294.
41 Daniels and Dayvault (2006), 294.

cone found in the cliffs of the Isle of Portland, identified in 1866 as *Araucaria sphaerocarpa*[42] of the family Araucariaceae. Another early find was a leafy conifer shoot from just above the Great Dirt Bed, illustrated in 1884 and tentatively assigned to the foliage morphogenus *Cupressinocladus*.[43] Small scale-like leaves of this type can be preserved occasionally in Purbeck strata as they are adpressed to the shoot, as for example are the leaves of today's controversial garden hedging tree x*Cupressocyparis leylandii* of the Cupressaceae family (the 'x' indicating that it is a cross). A morphogenus is a unit of classification, or taxonomic unit, assigned to isolated fossil plant parts such as leaves, seeds or cones with defined common characteristics, and possibly representing one stage of development. It permits the allocation of scientific names to disarticulated parts that can seldom be incorporated into whole plant reconstructions, and indicates 'structural similarity but not necessarily biological relationship'.[44] Until recently the term 'form-genus' was commonly used for a taxonomic unit that cannot readily be attributed to one family, and as is the case with morphogenera, some form genera served 'no other purpose than a way of reporting fragmented fossil material'.[45]

The extensive exposures of the Purbeck Fossil Forest in Dorset inevitably concentrated interest in that county, but exposures in the Vale of Wardour in Wiltshire were also recorded in the nineteenth and early twentieth centuries. In 1895 a horizon from the lower part of what is now the Purbeck Limestone Group at Chilmark Ravine, site of quarrying operations that date back to medieval and possibly Roman times, was described as follows: 'Dirt Bed, 1 ft. to 18 ins. thick, like the Great Dirt Bed of Portland, a carbonaceous clay with the remains of Cycads, and with rounded lumps of limestone and

---

42   Francis (1983), 278.
43   Francis (1983), 278.
44   Taylor, Taylor and Krings (2009), 1040.
45   Stewart and Rothwell (1993), 30.

Fossil wood collected in 2000 from a temporary exposure of the basal Purbeck Limestone Group beds in the Nadder valley, Wiltshire. Finds of fossil wood including tree stumps and bennettites date back at least to the nineteenth century in south-west Wiltshire, and pieces are still found from time to time in ploughed fields. Specimen: Author's collection. Photo: Author.

decomposed chert.'[46] In this same bed there was recorded 'an upright and rooted stump of a tree, the stem standing about 6 feet high.'[47] Specifically identified in Chilmark Ravine was the bennettite *Mantellia (Cycadeoidea) microphylla*, Buckland's *Cycadeoidea microphylla*. To the west of Ridge, a hamlet within the parish of Chilmark, reference was made in 1903 to the Purbeck basement bed as 'a tufaceous limestone with large masses of wood.'[48]

Wiltshire was a backwater and the focus for discoveries remained in Dorset. During the great nineteenth-century age of quarrying, when quarries on the Isle of Portland produced vast quantities of stone for shipment to London, finds were made that seemed unlikely to be repeated. Quarrying operations subsequently reduced in scale

46 Cited Reid (1903), 21.
47 In Reid (1903), 21.
48 Reid (1903), 23.

and became ever more mechanised, and some of the more interesting areas became worked out. A few years from now mining will have taken over from quarrying in a good proportion of the remaining Isle of Portland stone extraction industry. Despite this, a number of significant fossil finds were made and studies carried out during the latter half of the twentieth century.

In the late 1950s studies were undertaken of miospore assemblages from what were still known as the Purbeck Beds.[49] Miospores are small spores and pollen grains of a size below 200µ or micrometers, and results showed that up to 70 per cent of the contents of these assemblages consisted of pollen known as *Classopolis*. Three species from this morphogenus of coniferous origin were identified by G. Norris in 1969.[50] Other pollen grains were attributed to other conifers and to bennettites, and spores of ferns were also found. Meanwhile a rare assemblage of plant macrofossils – fossils large enough to be seen clearly with the naked eye – was recovered from the Purbeck basal beds at Portesham Quarry near Weymouth in Dorset. These finds were published in a 1975 paper[51] and, in what was a remarkable leap forward by Purbeck standards, eight new species of land plant, or rather parts of land plant, were described.

Portesham Quarry is the site of the most westerly exposure of the Purbeck Dirt Bed horizons, and the 1975 paper examined finds from a bed of cherty limestone. This bed, named the Charophyte Chert by Ian West in 1961, lay at the base of a lagoonal clay seen as the lateral equivalent of the Great Dirt Bed. Chert is a hard microcrystalline rock composed, like flint and the petrified wood of the Fossil Forest, of silica. Unlike flint it tends to form in lenses and layers as opposed to irregular nodules. Charophytes are predominantly freshwater algae, preserved in this case in silica along with the fossil land plants. Fossils

---

49 Francis (1983), 280.
50 Francis (1983), 289.
51 Barker, Brown, Bugg and Costin (1975).

were extracted from the friable cherty limestone through treatment with hydrochloric acid and sent to Reading University, where the renowned palaeobotanist Tom M. Harris supervised further work by co-authors Misses Brown, Bugg and Costin.[52] There could have been few better supervisors – during a long and distinguished career Harris had, among numerous other works, produced a landmark five volume work on the Middle Jurassic plant fossils of the Cleveland Basin of Yorkshire, gaining a worldwide reputation as a researcher.

The land plants were described by Brown and Bugg under three subdivisions. The first included a single species of horsetail of the Order Equisetales, not unlike the horsetails that can be found growing to this day. The small and fragmentary specimens of stems were attributed to a previously described species *Equisetum mobergii* from rocks of approximately similar age – Upper Jurassic to Lower Cretaceous – in Sweden. The match was not perfect, although this was in part because the Swedish specimens had been preserved through compression rather than three-dimensional silicification.[53]

The second section on land plants dealt with conifer remains, and included one twig bearing spirally arranged scale-like leaves assigned to the foliage morphogenus *Brachyphyllum*. Owing to its poor condition this specimen was given no species name and hence referred to as *Brachyphyllum* sp. Also described by Brown and Bugg were seven new species of unclassified conifer seed assigned to the morphogenus *Carpolithes*, a genus that incorporates numerous seed species of uncertain origin and relationships, and two incomplete seeds that were not given a species name. Of the seven named species only two could be compared by the authors with previously described material. In the case of one of these, *Carpolithes westi*, the situation was a little ambiguous. According to Brown and Bugg, 'These seeds

---

52   Barker, Brown, Bugg and Costin (1975), 419–21.
53   Brown and Bugg (1975), 428.

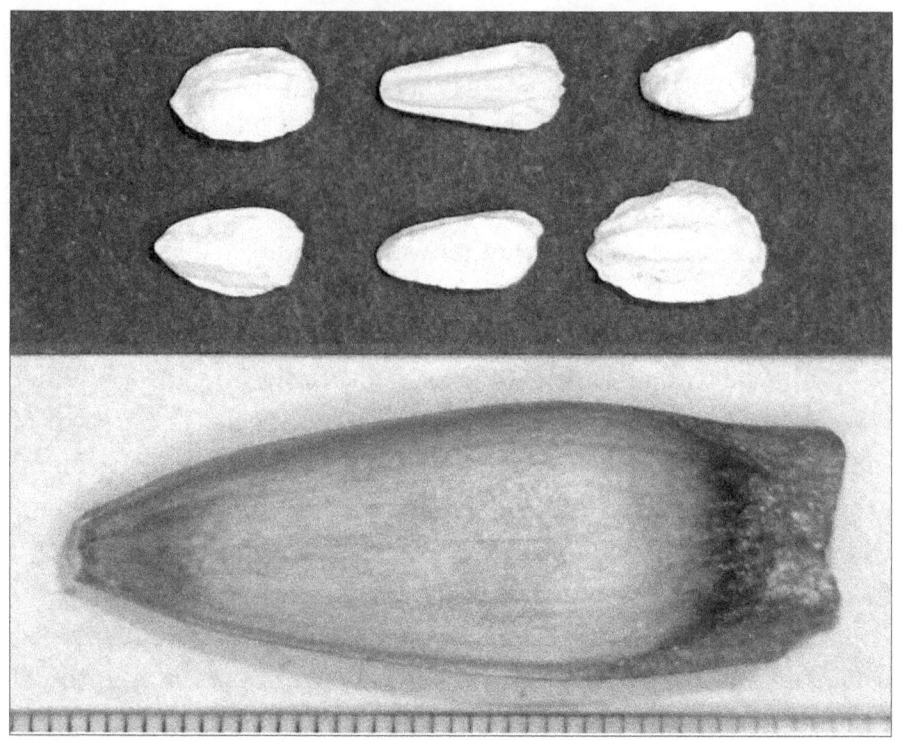

Seeds or possible araucarian cone scales (top) collected around 2006 from Wiltshire's plant and reptile bed and attributable to the species Carpolithes westi, first discovered in England in the basal Purbeck Limestone Group beds of Portesham Quarry, Dorset. The large specimen (bottom) is a present-day araucarian cone scale from the monkey-puzzle tree Araucaria araucana. Specimens: Author's collection. Photo: Author.

look like those figured by Seward 1904 ... as "Araucarites" from the English Inferior Oolite and *Carpolithus Lindleyanus* by Phillips 1871 ... *C. westi* might be the remains of an *Araucarites* seed, perhaps somewhat water-worn."[54] The features that marked out *Carpolithes westi* as a distinct new species were thus a little vague.

The third section of Brown and Bugg's description of land plants from the Charophyte Chert dealt with two specimens definitively attributed to the Araucariaceae. They were cone fragments, possibly

---

54   Brown and Bugg (1975), 434.

from the same cone, and were described as a new species *Araucarites sizerae*. The possibility that *Carpolithes westi* could in fact be part of a mature cone scale from this newly described species was considered, giving an idea of the difficulties involved in interpreting such disarticulated fossil plant remains. Despite some discernible differences, the *Araucarites sizerae* cone fragments were seen as most resembling among other known fossils the species *Araucaria mirabilis* from the famous Cerro Cuadrado Petrified Forest of Argentinian Patagonia.[55]

These new land plants from Portesham Quarry, found in association with fossil wood specimens that included a 15ft petrified tree trunk coated in tufa or porous limestone, originally thought to be a fossil elephant, expanded knowledge of the Purbeck Fossil Forest while simultaneously creating a new set of questions. Horsetails and a new araucarian conifer contributed to a sense of some identifiable diversity in the flora, but whether they had all grown within the main Purbeck Fossil Forest environment to the east could not be ascertained. The seeds represented further evidence of floral diversity, but beyond speculation over *Carpolithes westi* they could not be associated with any other known plant parts.

In the early 1980s Jane E. Francis engaged in extensive studies of the Purbeck Fossil Forest, and the results of her work were presented in two papers. The first, on the dominant tree of the forest, was published in 1983[56] and the second, on the forest environment, followed in 1984.[57] As a result new illustrated reconstructions were made, and the works of Francis were credited with stimulating an increase in interest in the Purbeck Fossil Forest.[58]

Central to these works was a detailed study of the fossil conifer

---

55 Brown and Bugg (1975), 434–5.
56 Francis (1983).
57 Francis (1984).
58 Cleal, Thomas and Batten (2001), 99.

wood. The process of silicification often preserves anatomical detail, providing a means of identification for the experienced professional. The most clearly visible features of a transverse cross section of such fossil wood are growth rings and rays radiating from the centre. With a fairly low level of magnification it is sometimes possible to see the walls of individual cells exposed on the surfaces of clean breaks. Where visible, this feature enables differentiation between angiosperm and conifer wood, but detailed identification to a wood morphogenus typically requires cutting and thin-sectioning to produce translucent slides for examination under powerful microscopes. Three sections cut along mutually perpendicular planes are needed.

Following an examination of many samples of Purbeck Fossil Forest wood, Francis was able to attribute over 90 per cent of them to the wood morphogenus *Protocupressinoxylon*. This lengthy name breaks down into three components: *proto* means early or primitive, *cupressin* derives from *Cupressus*, the genus name of the extant cypress trees, and *-oxylon* is the standard suffix for fossil wood genera based on the Greek word for wood. *Protocupressinoxylon* is thus 'early cypress wood', the name deriving in part from recognisable anatomical cupressoid or cypress-like features. Not all of its features link this wood to the cypresses, however, and the genus is assigned to the extinct conifer family Cheirolepidiaceae.

Several species of fossil wood from localities as diverse as the Isle of Wight and Malaysia had already been attributed to the genus *Protocupressinoxylon*. None of these matched certain features of the Purbeck Fossil Forest wood studied by Francis, and she was thus able to assign it to a new species, *Protocupressinoxylon purbeckensis*. From this she moved on to consider the likely foliage, cones and pollen that would enable a whole plant reconstruction. As none of these had ever been found directly attached to fossil wood, she depended on association. The process is not without its dangers, and according to one expert opinion, 'Experience has shown that the association of

The broken end of a 2.5cm diameter piece of fossil wood, probably root, from south west Wiltshire's Fossil Forest. Anatomical details are preserved, including growth rings and radiating cell rows. Further study requires the preparation of three thin microscope sections, which with good preservation can lead to identification to a morphogenus. Specimen: Author's collection. Photo: Steve Clifford.

dispersed organs is not, on its own, proof that they were once part of a single plant.'[59] Several factors can come into play here, including the selective dismemberment, decay and transport of plant material.

Compressed fossil foliage in the form of scaly leafed shoots comparable to those of the extant cypress family was found in limestone overlying the dirt beds, and in the Great Dirt Bed itself at God Nore on the Isle of Portland. The significance of the association with fossil wood was tempered by the fact that the bed here is 'reworked

---

59 Cleal and Thomas (2001), 7.

and resedimented', the fragmentary material only occurring on 'a few organic-rich laminae'.[60] As with the leafy shoot described in 1884, the shoots were attributed to the morphogenus *Cupressinocladus*. On the basis of evidence from a number of specimens they were further interpreted as belonging to the species *Cupressinocladus valdensis*, originally described from the Lower Cretaceous English Wealden deposits. The scattered carbonised shoots in the Great Dirt Bed were found in association with small carbonised male cones not directly attached to the foliage, and attributed by Francis to the cone genus *Classostrobus*. Lacking in internal structure, the cones were too poorly preserved to be allocated to a species. According to Francis, the close association of shoots and cones along with similarities in cuticles or outer protective membranes 'strongly suggests that they are part of the same plant'.[61] The pollen sacs were missing from the *Classostrobus* cones, but isolated grains were found adhering to the cuticles. Francis identified these as being of *Classopolis* type, but not specifically assignable to any described species.

The frequency and association of the wood *Protocupressinoxylon purbeckensis*, the foliage *Cupressinocladus valdensis*, the cone *Classostrobus* sp. and the pollen *Classopolis* sp., along with the scarcity of other wood and foliage, led Francis to conclude that 'the wood, shoots, male cones, and pollen described here are considered to represent parts of one conifer which dominated the Lower Purbeck forests'.[62] Having dealt with the technicalities, Francis approached the million dollar question: what did these trees actually look like? Using illustrated examples that included a reconstructed silicified trunk from the Isle of Portland and a stump with 1m of trunk from Chalbury Camp in Dorset, some evidence as to the shape of the base of the tree was presented. Roots radiated out from thickened and slightly

60   Francis (1983), 279.
61   Francis (1983), 288.
62   Francis (1983), 289.

twisted stumps up to and over 1m in diameter. The roots appear to have been unable to penetrate the underlying limestone, and on contact with it to have spread horizontally. From the stump the trunk tapered upwards, in the case of the Isle of Portland specimen to its total preserved height of about 2m. The Chalbury Camp specimen also had a branch 42cm long attached to the trunk, subtending at an angle of about 40 degrees some 44cm above soil level.[63] This was a rare find, and along with knots marking positions of other branches it could be seen that branching occurred on an irregular basis, with several branches arising from near the base. The key to the whole reconstruction, however, was an interpretation that the tree 'probably had a more or less monopodial growth form'.[64]

Monopodial trees are those with a single trunk reaching to the apex, typically represented by the traditional Christmas tree, the Norway spruce *Picea abies*. The illustrated reconstruction of the dominant Purbeck conifer produced by Francis[65] thus had a single tapering trunk, but differed from the Norway spruce in that the branches were sparser and arranged irregularly rather than in whorls. The taper from a diameter of about 1m at the base of the trunk to the apex about 8m above soil level was far more pronounced, leading to a squat, heavy appearance.

This basic 1983 reconstruction has made its appearance many times since. In 2001 it appeared twice under different guises in *Mesozoic and Tertiary Palaeobotany of Great Britain*. In the first instance it was illustrated as a 'Reconstruction of a cheirolepidiaceaen conifer, a tall forest tree' from the Jurassic,[66] redrawn from the illustration in Francis's paper with a more pronounced twist at the base of the trunk, a few extra minor branches and considerably more

63 Francis (1983), 279, 290.
64 Francis (1983), 290.
65 Francis (1983), 289.
66 Cleal, Thomas and Batten (2001), 103.

foliage. In the second instance, taken from a 1996 paper by J. Watson and K.L. Alvin, it was illustrated as '*Cupressinocladus valdensis* ... a tall forest tree'[67] from the Lower Cretaceous Wealden flora. In this case the drawing was all but identical to the 1983 original. The identification of the Purbeck foliage as *Cupressinocladus valdensis* on the basis of limited material had led, by corollary, to the identification of foliage from the Lower Cretaceous as belonging to the same species of tree, despite there being likely climatic and geographical differences between the Purbeck Fossil Forest environment and the Wealden environment. If the trees *were* the same, the name applied to the whole plant had thus become based upon different parts – Upper Jurassic Purbeck trees were 'the cheirolepidiaceous conifer

A fallen monopodial mountain pine Pinus mugo decays in the Eastern Pyrenees. It grew on the edge of a closed conifer forest, such as could have grown in semi-arid lowland regions of England during Upper Jurassic times. Exceptions to monopodial growth, probably resulting from damage, are not unusual among these trees. Photo: Author.

67   Thomas and Batten (2001b), 142.

Protocupressinoxylon purbeckensis'[68] based on the name of the wood, whereas Lower Cretaceous Wealden trees were *Cupressinocladus valdensis* based on the name of the foliage. One reference leads to another, and in referring to the Cheirolepidiaceae it is stated somewhat confusingly in the most recent exhaustive textbook on palaeobotany that '*Cupressinocladus valdensis (Protocupressinoxylon)* is a Late Jurassic member of the family that was the dominant tree in southern England during the Wealden (Francis 1983)'.[69]

Not surprisingly, even professional palaeobotanists can state that 'Non-paleobotanists may find the nomenclature used in paleobotany confusing and perhaps cumbersome'. Correct procedure on reconstruction of an entire plant[70] is to provide a new name (in this case the species name *purbeckensis* is new), to base the generic name on the first part to have been formally named, or to use an informal name such as 'the *"Protocupressinoxylon"* tree'. Whatever the procedure adopted, a piecemeal approach can confuse rather than clarify plant reconstructions.

The reconstruction by Francis had in fact, in her own words, involved 'an imaginative interpretation of the data', and indeed such interpretations of scientific evidence are essential for breathing life into palaeontology. Where there was a lack of direct evidence 'comparisons have been made with modern conifers from semi-arid regions, such as *Juniperus oxycedrus* Linnaeus and *Cupressus macrocarpa* Hartweg'.[71] Although climatic conditions under which these juniper and cypress species grow are possibly comparable to the conditions in which the Purbeck Fossil Forest trees had grown, and although the trees *could* have been comparable, they belong to the Cupressaceae and the relationships here are complex and not clear-cut. There are indirect

---

68 Cleal, Thomas, and Batten, (2001), 110
69 Taylor, Taylor and Krings (2009), 833.
70 Taylor, Taylor and Krings (2009), 42.
71 Francis (1983), 291.

links – as has been seen in the introduction, the Cheirolepidiaceae have been interpreted as 'transitional taxodioids' and members of the Taxodiaceae have been incorporated into the Cupressaceae in recent classifications. Fossil wood identifications are not without problems, however, and even the validity of the morphogenus *Protocupressinoxylon* has recently been brought into question – in 2008 it was claimed to be 'illegitimate' by a French palaeobotanist.[72] The very names *Protocupressinoxylon* and *Cupressinocladus* could also tend to mislead because in their etymology they 'imply a relationship with the Cupressaceae',[73] and despite representing fossil plants attributed to the extinct Cheirolepidiaceae such names could only serve to reinforce the idea of the Purbeck Fossil Forest as consisting of cypress-like trees. The interpretation survives to this day largely unquestioned, with a local geologist recently reported as describing a 3 ton Purbeck tree stump found at Poole in Dorset as coming from a 'pine tree, similar to a cypress or juniper'.[74] This could be the case, but is far from reliably established on the basis of an imaginative interpretation of fragmentary data.

Francis noted that in the early to mid-nineteenth century observations had been made of straight trunks from the Purbeck Fossil Forest up to 13m in length,[75] indicating far taller trees than the approximately 8m reconstruction she illustrated. She further stated that the tree trunk illustrated by Fitton in 1836 showed a taper of 9.5cm from a diameter of 47.8cm at the base to a diameter of 38.3cm at a height of 5.51m, concluding that 'With monopodial axes of this length trees of over 20m can be envisaged'.[76] This conclusion has to

---

72   Ian West *Purbeck Bibliography* www.soton.ac.uk/~imw/Portland-Isle-Geological-Introduction.htm (visited 4 April 2010).
73   Batten (2002), 16.
74   Bournemouth *Daily Echo*, 11 February 2009.
75   Francis (1983), 290.
76   Francis (1983), 290–291.

be set against the fact that Fitton's stump and trunk is a very different tree to her own reconstruction. Similar in basic shape to another specimen housed in the Dorset County Museum in Dorchester, it has a gentle taper, no twisting at the base of the trunk and a division of the trunk into two branches about 1m from the truncated top. Although the branches are not exactly equal in size, the tree is not apparently monopodial.

Some later reconstructions of the Purbeck Fossil Forest have used Francis's work as a baseline from which modifications have been made. An illustration in the official Jurassic Coast guide[77] shows trees in which the severe taper and rigid straightness of the trunk

Foliage of the Leyland Cypress 'Leylandii' of the Family Cupressaceae. Impressions of fundamentally similar scale-like foliage adpressed to shoots have been found in the lower beds of the Purbeck Limestone Group of Dorset and incorporated into the reconstructed Fossil Forest tree Protocupressinoxylon purbeckensis. Photo: Author.

have been modified, producing taller forms with larger branches and slightly meandering trunks.

77 Brunsden (ed.) (2003), 51.

Although some 90 per cent of the fossil wood sampled was identified by Francis as belonging to *Protocupressinoxylon purbeckensis*, two other wood morphogenera were recognised. Distinctive araucarian-type anatomical detail led to some samples from the Isle of Portland and Lulworth being attributed to the genus *Araucarioxylon*, supplementing cone evidence from the Isle of Portland and Portesham Quarry as to the presence of araucarian trees inevitably associated with today's monkey puzzle tree. One silicified trunk, probably from the Lower Dirt Bed, was attributed to the conifer morphogenus *Circoporoxylon*,[78] also anomalously stated to be 'only represented by wood fragments'.[79] A recently described species of *Circoporoxylon* from Argentina was attributed to the family Podocarpaceae,[80] and early podocarps can thus be seen as a possible component of the Purbeck Fossil Forest.

The reconstruction of whole plants is only one part of modern palaeobotany, and Francis considered a wide range of related questions. How old were the trees? What was the density of the forest in which they lived? What was the climatic environment in which they grew? How did they come to be preserved?

In terms of age, the Purbeck Fossil Forest of Dorset was judged on the basis of growth rings to consist of mature trees with a probable age range of 200 to 700 years.[81] The growth rings also provided important evidence of a seasonal environment, as expounded by Francis in 1984: 'The narrow and variable growth rings of the trees indicate that conditions were marginal for tree growth and highly irregular from year to year. Comparison with modern tree-ring data suggests that the Purbeck climate was of Mediterranean type, with

---

78 Francis (1983), 288.
79 Francis (1984), 289.
80 Gnaedinger (2007), 77.
81 Francis (1983), 291.

warm winters ... but with hot, arid summers.'[82] A 'probably sparse undergrowth' had already been conjectured,[83] and the overall image was a far cry from that offered by Barker and his co-authors in 1975: 'Thus we have a vivid picture of a luxuriant coastal swamp, thickly forested and teeming with life'.[84] This latter image could well have been shaped in part at least by the discovery of horsetail fragments at Portesham Quarry, often associated with damp or marshy conditions, and although horsetails occupy a broad range of habitats, such plants could appear incongruous along the semi-arid shore of a hypersaline gulf as postulated by Francis.[85] The presence of horsetails has tended to be neglected – they were still described as 'absent' from the Jurassic vegetation of southern England in a 2001 appraisal.[86]

One of the advantages of studying *in situ* fossil forests is that the tree density can sometimes be calculated. In 1884 Robert Damon had noted the discovery of seven tree stumps and two cycadophyte trunks within 'a few square yards',[87] but a detailed study was lacking. In the Purbeck Fossil Forest the stumps are either preserved or, in many cases, their original locations marked by stromatolitic limestone precipitated by algal or bacterial growth following submergence. This formed domes or burrs over many stumps that subsequently rotted away, often leaving fine moulds on which the impression of the wood grain can be seen.[88] Through mapping at four localities, Francis calculated forest densities ranging from one tree per 15sq.m to one tree per 54sq.m, commenting that 'These observations suggest closed

82   Francis (1984), 285.
83   Francis (1983), 292.
84   Barker, Brown, Bugg and Costin (1975), 420.
85   Francis (1983), 292.
86   Cleal, Thomas and Batten (2001), 101.
87   Cited Francis (1983), 278.
88   Francis (1984), 289.

forest conditions, which is perhaps surprising in such a semi-arid climate.'[89]

Apparent anomalies between the flora and certain climate indicators are also reflected in the sediments. The Portesham Quarry exposure of the Great Dirt Bed horizon, according to Francis, 'represents a marginal lagoonal clay which contains an anomalous association of both silicified wood, conifer shoots, seeds and charophytes (Barker et al., 1975) along with nodules of silicified evaporite pseudomorphs'.[90] Charophytes, as noted earlier, are predominantly freshwater algae, whereas evaporites are rocks precipitated from evaporating water as salinity increases. An example of a lagoon in South Australia, however, was used by Francis to demonstrate that freshwater conditions along the margin of a lagoon can co-exist with evaporite formation at the centre of the lagoon.

The complexities involved in interpreting a past environment become clear in any study of the Purbeck Fossil Forest. The vegetation shared some general similarities with the earlier Middle Jurassic Yorkshire flora – macrofossils indicate the common occurrence of three conifer families, bennettites and horsetails – but the climate had apparently become drier and, according to growth ring evidence, seasonally erratic. The miospore assemblage from the Great Dirt Bed indicates some ferns and lycopods,[91] which along with the horsetails from Portesham Quarry and the closed nature of the forest suggest a possibly mesic or moderately moist forest floor. Against this, charcoal fragments indicate a susceptibility to forest fires, and evaporite formation has been interpreted as a clear indication of growing Upper Jurassic aridity. The Russian palaeobotanist V.A. Vakhrameev mapped the development of such climatic zones in both the northern and southern hemispheres. He concluded that in Upper Jurassic times

---

89   Francis (1984), 290.
90   Francis (1984), 298.
91   Francis (1984), 289.

'dramatic aridization of climate took place within a broad belt crossing south England, France, Switzerland, western Ukraine, the Caucasus, Middle Asia, Mongolia and western China. The whole broad belt is noted for sedimentation of evaporates in disjunct basins.'[92] At a time when the Atlantic Ocean was just beginning to open up, and the continental masses of the Americas, Eurasia and Africa were in an early stage of separation as the former super-continent of Pangea began to break apart, a vast semi-arid to arid region extended from

Horsetails growing along the valley of a tributary of the River Nadder in Wiltshire. Related to tree-sized ancestors from the Carboniferous period, these plants with their distinctive leaf whorls have changed little since Jurassic times, and evidence from Portesham Quarry in Dorset indicates their presence in the margins of the Purbeck Fossil Forest environment. Photo: Author.

92 Vakhrameev (1991), 47.

the southern USA to southern Argentina and from central Europe and China to South Africa.[93]

Francis identified the process of silicification in the Purbeck Fossil Forest wood as a further indication of an evaporitic environment and semi-arid climate. In the first phase of this process silica has to be dissolved in water, originating either from volcanic ash, from material of organic origin such as the spicules that make up the delicate skeletons of sponges, or, as suggested by Francis in the conditions of the Purbeck environment, from detrital quartz and silicate clay minerals. This dissolution occurs during seasonally high pH or acidic conditions, and in lower pH conditions the silica is precipitated in areas such as those where plant material is decaying.[94] A specific attraction of dissolved silica towards wood and a complex templating process are now recognised. Initially silicic acid molecules penetrate woody tissue and bond with vascular woody tissue.[95] The silica then develops through intermediate opal stages to microcrystalline chalcedony. Francis described the Purbeck wood as 'petrified by quartzine', 'a type of length-slow chalcedony often associated with evaporitic environments'.[96] Length-slow chalcedony is a variety of fibrous quartz.

Most trees, of course, do not fall down and become silicified. They rot away and are recycled. Fossilisation is an extremely rare occurrence compared with the number of trees that have grown on this planet, and in the case of the Purbeck Fossil Forest it is believed that the land was swamped by rising hypersaline lagoonal water that both drowned and preserved the trees until the process of silicification began. Even under the relatively favourable conditions for preservation that clearly existed, many of the stumps and branches rotted away to leave those voids in the overlying stomatolitic limestone that enabled

---

[93] Vakhrameev (1991), 255.
[94] Francis (1984), 299.
[95] Daniels and Dayvault (2006), 179.
[96] Francis (1984), 299.

the forest density to be estimated.

It is indicative of the nature of the fossil record that the Purbeck Fossil Forest has been described in recent years as 'the most complete fossil record of a Jurassic forest in the world',[97] and yet in a 2002 paper it was stated that apart from conifer and bennettite stumps and trunks 'the plant macrofossil record for the Purbeck succession is very limited'. As for the hinterland 'mosaic of [plant] communities', the subject 'is at present almost entirely a matter for speculation'.[98]

No one ever knows what new discoveries lie around the corner, of course, and even since that 2002 appraisal new finds have been made in Dorset, including fallen *in situ* trees at Landers Quarry from the middle part of the Purbeck Limestone Group. These largely carbonised and unrecoverable specimens, with a small amount of silicification, were described in 2010 as the most significant finds since Lulworth and interpreted as having been killed by flooding.[99] Work on the Weymouth Relief Road in 2009 resulted in the collection of plant material from a previously recorded plant bed in the Cherty Freshwater or Marly Freshwater Member of the Purbeck Limestone Group, with finds including carbonised roots, branching plants, a conifer shoot and possible fern pinnules.[100]

Whether the Purbeck Fossil Forest grew along the semi-arid shores of a hypersaline lagoon or on a luxuriant coastal swamp, the number of described macrofossil plant species clearly represents no more than a small proportion of the total vegetation. The animal life of the Purbeck Limestone Group is more broadly represented, partly in the form of invertebrates such as small crustaceans and molluscs

---

97 Brunsden (ed.) (2003), 50.
98 Batten (2002), 18–19.
99 Natural Stone Specialist *Tree fossil find on Purbeck most significant since Lulworth* www.naturalstonespecialist.com (visited 31 January 2011)
100 Ian West *Purbeck Bibliography* www.soton.ac.uk/~imw/Ridgeway-Railway-Cutting.htm (visited 31 January 2011)

that act as important aquatic environmental indicators. Insect fossils include the remains of beetles, wasps, caddis flies, lacewings, dragonflies and, from Teffont Evias in south-west Wiltshire, an exceptionally preserved fly *Obliogaster fittoni* on a piece of limestone.[101] A fine selection of insect remains was collected by Robert Coram in 2009 from the temporary Weymouth Relief Road construction site.[102] Most significant, however, is the wide range of vertebrates. The Purbeck exposures at Durlston Bay in Dorset have produced one of the most diverse fossil vertebrate faunas from Britain, including over forty species of reptile as well as twenty species of early mammal.[103] An evaluation of vertebrate diversity within the Purbeck Limestone Group as a whole reported species totals, excluding fish, of 'four amphibians, four turtles, 13 lepidosaurs, seven crocodiles, three pterosaurs, five dinosaurs, two marine reptiles and 28 mammals'.[104] Lepidosaurs are reptiles with overlapping scales belonging to the two orders Sphenodontia and Squamata (see Table 3). Despite this abundance of evidence, however, there are basic problems in reconstructing the animal life of the Purbeck Fossil Forest environment. Firstly, the vertebrate remains are mostly found in higher beds, many of them in the upper beds of the Lulworth Formation, whereas the Fossil Forest lies near the base. Those from above the Lulworth Formation, i.e. from the Durlston Formation, come from a period when evidence suggests 'the climate became progressively more humid'.[105] Of the nineteenth-century finds, no record was kept of the horizons from which many were collected. Secondly, in contrast to the largely invertebrate lagoonal creatures, 'it is more difficult to determine what animal life

---

101 McBain and Nelson (2003), 9.
102 Ian West *Purbeck Bibliography* www.soton.ac.uk/~imw/Ridgeway-Railway-Cutting.htm (visited 31 January 2011)
103 Benton, Hooker and Cook (2005), 59.
104 Milner and Batten (2002), 5–6.
105 Batten (2002), 13.

was like on land because the remains of vertebrates consist mainly of teeth and very small pieces of bone'. Along with dinosaur footprints they provide no more than 'a tantalizing glimpse of a diverse fauna'.[106] Although tantalising, it has been the best evidence so far available for

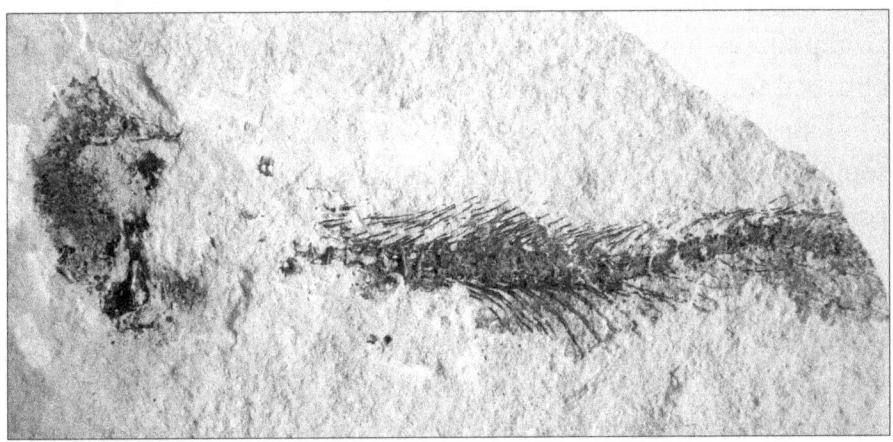

A fossil fish from the base of the Purbeck Limestone Group, south west Wiltshire. Records of abundant fish remains from this horizon date back to the nineteenth century. The limestone in which they occur lies immediately above the marine Portland Stone Formation and below the terrestrial Dirt Beds, representing intermediate freshwater to brackish conditions .
Specimen: Author's collection. Photo: Author.

attempting a reconstruction of the world of the Purbeck Fossil Forest and is thus worthy of a brief description.

Study of the vertebrate Purbeck fauna has centred on sites in Dorset, largely along the coastal exposures and quarries of the Isle of Purbeck. Wiltshire does occasionally feature in the literature, however. The fossil fish, although generally seen as beyond the scope of this book, deserve a brief mention as they probably lived in water along the forest margins. At the old Oakley Quarry near Tisbury, Henry W. Bristow observed of the junction between the Portland Stone and the

---

106 Batten (2002), 13.

Purbeck Beds that 'The line of demarcation is crowded with fish.'[107] Thin limestone beds with fossil fish remains, representing freshwater or brackish conditions between the underlying marine limestones and the overlying palaeosols of the Purbeck Fossil Forest, are found elsewhere in the Vale of Wardour. Higher in the Purbeck Limestone Group succession, some of the best preserved Purbeck fish 'have been obtained from the neighbourhood of Teffont Evias'.[108]

Bridging the freshwater and terrestrial environments are some rare Purbeck amphibian fossils. The first unconfirmed report concerns frog remains from Swindon, Wiltshire, mentioned by W.H. Hudleston in 1876,[109] but it was not until over a century later that confirmed amphibian remains were discovered in the Cherty Freshwater Member of Dorset's Lulworth Formation. They were found at Sunnydown Farm Quarry on the Isle of Purbeck and reported by Paul Ensom in 1988.[110] Further finds were made, and by 2002 the fragmentary remains of two types of salamander, a primitive frog, and a salamander-like amphibian from a now extinct group had been described.[111] Whether such creatures would have lived on the margins of the Purbeck Fossil Forest depends, of course, upon the year on year availability of freshwater in which to breed – or, perhaps, on specialised burrowing and reproductive adaptations to survive droughts.

Of the reptiles, turtle remains are relatively abundant from several horizons. Classification of these was revised in 2004, with three genera recognised on the basis of shells, *Pleurosternon*, '*Glyptops*' and *Helochelydra*, and one on the basis of the skull, *Dorsetochelys*.[112]

107 Cited Reid (1903), 22.
108 Reid (1903), 18.
109 Hudleston (1876).
110 Ensom, (1988), 148–50.
111 Evans and McGowan (2002), 103.
112 Milner (2004), 1441.

Along with most of today's turtles, these all belong to sub-order Cryptodira in which the head retracts back into the shell. The thirteen described species of lepidosaur have mostly been found in the Cherty Freshwater Member and include the lizard genus *Dorsetisaurus* of the order Sqamata. The crocodiles include large types such as the 'Swanage crocodile' *Goniopholis* of the family Goniophilidae, and a small species of the genus *Theriosuchus*, the remains of which have been interpreted as belonging to dwarf crocodiles around 18in long[113] – they are not technically dwarf, and it has been 'suggested that the "stigma" of dwarfism should be removed from them'.[114] *Theriosuchus* could have been insectivorous and belongs to the Atoposauridae, a family that was possibly ancestral to today's crocodiles.[115] Button-like crushing teeth have also been found in the Purbeck Limestone Group of Dorset, possibly belonging to the small crocodile genus *Bernissartia* and cautiously designated as cf. *Bernissartia*.[116] Such teeth are frequently associated with animals that feed on shellfish. The best-known beds yielding crocodile remains are in the Lulworth and Durlston Formations, formerly Middle and Upper Purbeck Beds, at Durlston Bay, although abundant fossils including crocodile and turtle remains are recorded from Purbeck marls at Teffont Evias.[117]

Determinate fossils of flying reptiles or pterosaurs from the Purbeck Limestone Group are rare and the exact beds and locations of early finds poorly recorded. The known material was revised in 1995 and three genera belonging to two families recorded.[118] Two genera of ctenochasmatids, *Gnathosaurus* and *Plataleorhynchus*, are represented by pterosaurs that probably fed by sieving, filtering and raking through

113 Reid (1903), 18.
114 Joffe, J. (1967), 638.
115 Joffe, J. (1967), 629.
116 Salisbury (2002), 138.
117 Geddes (2000), 202.
118 Howse and Milner (1995), 73.

mud and water for small invertebrates. Along with the amphibians, the occurrence of such creatures 'further emphasises the freshwater-marginal source of much of the fauna'.[119] The third pterosaur genus, *Ornithocheirus*, is not represented by material determinate to species level and is described on the basis of fragmentary material. Other finds include isolated pterosaur teeth from Sunnydown Farm Quarry, and these reveal the problems that can be faced in identifying such fossils. Slender recurved teeth originally identified as representing the 'anterior fish-spearing teeth of small rhamphorhyncoids' were later alternatively interpreted as 'the slender recurved marginal teeth of a large ctenochasmatid'.[120]

The most spectacular of the dinosaurs, the lizard-hipped saurischians, are very poorly represented in the Purbeck Limestone Group of Dorset and even less so in Wiltshire. The only evidence of such herbivorous sauropods from these counties comes from footprints left in the mudflats before the soft lime mud through which they walked turned to stone. The preserved footprints were transferred by the reptiles' huge weight to layers beneath the original surface, complicating their interpretation. Dinosaur footprints and tracks are described in at least seven papers written or co-written by Paul Ensom between 1988 and 2008,[121] and in one case the recorded evidence of sauropod footprints comes from a horizon in the lower part of the Lulworth Formation that was formerly placed in the Lower Purbeck Beds. Such a find takes the dinosaur record down closer to the possibly more arid climate associated with the Purbeck Fossil Forest horizon.

More substantial fossil sauropod evidence has been found in Buckinghamshire. Purbeck conditions extended far beyond Wiltshire

---

119 Howse and Milner (1995), 87.
120 Howse and Milner (1995), 86.
121 Ian West *Purbeck Bibliography* www.soton.ac.uk/~imw/Purbeck-Bibliography.htm (visited 10 September 2009).

A broken sauropod tooth crown cf. Pelorosaurus from the gastropod micrite of the Portland Stone Formation, Vale of Wardour. The tooth has been worn down by the constant grinding of plant material. Teeth attributed to this genus have been reported from the Purbeck Limestone Group at Bugle Pit in Buckinghamshire. Identification: NHM. Specimen: Author's collection. Photo: Steve Clifford.

and Dorset, across much of southern England and over the Channel into northern Europe, and rocks of this age were formerly quarried around Thame and Aylesbury. In this area marine limestones of the Portland Stone Formation change upwards 'to lagoonal fine-grained limestones and marls of the Haddenham Formation', placed by the

British Geological Survey in the 'Purbeck Group'.[122] One quarry, Bugle Pit, is preserved as an SSSI although it is apparently in an unsafe condition. Fossil collecting dates back to the nineteenth century and 'the vertebrates recorded from this site include several sauropod teeth from the only known remains of *Pelorosaurus* sp. of Portlandian age in Europe.'[123] Portlandian rocks were laid down during the Tithonian age, and the assemblage of invertebrates known as ostracods at Bugle Pit is typical of other Portland-Purbeck marine to freshwater transitions.

Pelorosaurus was the first ever sauropod to be correctly named, on the basis of a 1.4m long humerus from the Lower Cretaceous Wealden deposits obtained by Mantell in 1849. The history of the genus, which is possibly a member of the brachiosaurid family, is confusing and various fragmentary material has since been attributed to it. However, the Bugle Pit teeth do confirm in conjunction with the Dorset footprints the widespread presence of large sauropods in southern England during deposition of the lower part of the Purbeck Limestone Group.

As for the carnivorous two-legged saurischians, the theropods, a trackway and some rare bones and teeth are recorded from the Purbeck Limestone Group of Dorset. Jaw specimens collected in the nineteenth century from Feather Quarry on the Isle of Purbeck, probably from the Cherty Freshwater Member, were described by Owen in 1854 as *Nuthetes destructor*. The remains were either from a small species of theropod or from a juvenile.[124] Isolated teeth were also found, closely comparable with those of dromaeosaurids, and '*Nuthetes* has also been reported from the Purbeck Limestone

---

122 BGS Lexicon of Named Rocks http://www.bgs.ac.uk/lexicon/lexicon/cfm?pub=HADD (visited 3 April 2010).

123 Bucks Earth Heritage Group http://www.bucksgeology.org.uk/sssi/bugle_pit.htm (visited 7 April 2010).

124 Milner (2002), 191.

of Wiltshire'.[125] Dromaeosaurids are raptors, probably feathered and renowned for their long slashing claws and fierce *Jurassic Park* reputations. Sizes ranged from under 2 to well over 4 and possibly 6m. One large Purbeck theropod tooth crown is closely comparable to teeth of the genus *Allosaurus*, represented by carnivores up to 12m long with two powerful legs and short arms.

Also of note is the discovery of teeth attributed to *Megalosaurus* in the Haddenham Formation at Bugle Pit in Buckinghamshire. *Megalosaurus* was the first scientifically described dinosaur, on the basis of Middle Jurassic material from Stonesfield in Oxfordshire. The description by Buckland was published in 1824, when the reptile was interpreted as a large lizard before the word 'dinosaur' had been coined. As with *Pelorosaurus* no complete skeleton has ever been found, and nineteenth-century teeth identifications have been the subject of various revisions. According to a recent study of the Purbeck theropods, the Buckinghamshire teeth share with the large *Allosaurus*-like tooth previously mentioned above 'characters in common with allosauroids', although ultimately they 'may not be determinate'.[126]

Remains of bird-hipped dinosaurs of Order Ornithischia are also rare in the Purbeck Limestone Group and the most significant fossil bones were found in Dorset during the nineteenth century. A jaw reported in 1874 by Sir Richard Owen, founder of the Natural History Museum in London, was originally named *Iguanodon hoggii* and is now assigned to the genus *Camptosaurus*. It is of some historical interest as it 'allowed the first definitive description of the pattern of tooth replacement of any herbivorous dinosaur'.[127] Probably a browser on low-growing vegetation, *Camptosaurus* was estimated to be around 3 to 5m in length, had large, strong hind legs and would have defended

125 Milner (2002), 191.
126 Milner (2002), 198.
127 Norman and Barrett (2002), 161.

itself through speed. The only other determinate ornithischian dinosaur from the Purbeck Limestone Group is *Echinodon becklesii*, described by Owen in 1861 from jaw and maxilla remains from either the Marly or the Cherty Freshwater Member of the Lulworth Formation at Durlston Bay. The animal from which these bones came was a possible heterodontosaurid, a small two-legged plant eater notable among the dinosaurs for its differentiated specialised teeth. Rare indeterminate remains of armoured ornithischian dinosaurs have also been recorded from the Purbeck Limestone Group, including a possible nodosaur tooth from Lulworth and a vertebra 'from the Dirt Bed of the Lower Purbeck Beds'[128] referred to the Ankylosauria – the Ankylosauria includes both nodosaurs and ankylosaurs. Protected by bony plates and possessing spines or bony tail-clubs, these dinosaurs were medium-sized herbivores with four relatively short legs. If the overall bone record for ornisthischian dinosaurs is poor, this may be due to lack of suitable conditions for the preservation of larger material, as their preserved footprints are comparatively abundant.

Fossils of marine reptiles are occasionally found in the Portland Stone Formation of Dorset, as would be expected in marine deposits, but are rare in the Purbeck Limestone Group where conditions would seldom have suited such creatures. An unusual example is a juvenile ichthyosaur at the Oxford University Museum of Natural History collected from an unknown location.

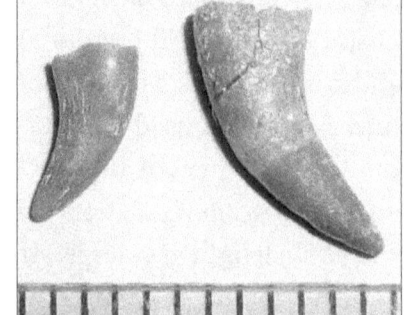

Small theropod teeth from Wiltshire's plant and reptile bed curve back in a similar manner to Nuthetes destructor teeth from Dorset. The dinosaurs from which they came were probably dromaeosaurs from the same sub-family as the velociraptors. The teeth were found through slow processing of a bulk sample of the bed. Specimens: Author's collection. Photo: Author.

128 Norman and Barrett (2002), 184.

Of all the Purbeck vertebrate remains, those of the mammals are regarded by many as the most significant. Fossil jaws of small shrew-to-rat sized creatures were first discovered in the 1850s and described by Owen in 1854, following which a large pit was excavated by Samuel Beckles in 1857 into the cliffs at Durlston Bay. These fossils from what was subsequently named the Mammal Bed, now in the Marly Freshwater Member, have been of major international importance in understanding mammals from the Jurassic-Cretaceous transition period. When put together with later finds, including teeth from Sunnydown Farm Quarry described by Paul Ensom and D. Sigogneau-Russell in various papers since 1994, the Purbeck mammal species list totalled twenty-eight by 2002.[129]

The early mammals included multituberculates with numerous grinding facets on their molars and large incisors. Of these probable seed eaters, the genus *Plagiaulax* has been recorded from Town Gardens Quarry at Swindon as well as from the type locality at Durlston Bay. The probably insectivorous technothere genus *Spalacotherium* has also been recorded from both localities. Other mammals from Durlston Bay include a docodont (a basal group of primarily herbivorous and insectivorous mammals), four possibly carnivorous triconodonts and six insectivorous cladotheres.[130]

The conditions under which small teeth and bones are preserved, typically through burial in sediments sorted and transported by moving water, are not the same as the conditions under which the footprints of dinosaurs walking across broad mudflats are preserved, which in turn are not the same as the conditions under which the stumps and trunks of forest trees are preserved *in situ* in a palaeosol. The question thus arises: in the complex network of changing Purbeck environments, laid down over geological time counted in

---

129 Milner and Batten (2002), 6.
130 Benton, Hooker and Cook (2005), 59.

millions of years, how plausible is it to construct composite pictures of the environment based upon evidence from different horizons? Even within the time spans of human history factors affecting some environments can change rapidly. Examples include climate changes and shifting water courses from which the supply of freshwater is essential for the survival of some elements of the flora and fauna.

Taking account of the above, it is possible to imagine a scenario in which, where there now stand the cliffs of Dorset's Jurassic Coast, the chalk downs of Dorset and Wiltshire, the hills and valleys of south Wiltshire's Vale of Wardour and the urban environs of Swindon, there lay a great area of mud flats, lagoons and flatlands along the shifting shores of an inland gulf. Conifers of a long extinct family dominated forests in which ancestral araucarians and podocarps and extinct bennettites also grew. A probably sparse undergrowth included some ferns and lycopods, possibly growing near standing water and watercourses. During hot Mediterranean to sub-tropical days reptiles ranging from small lizards to huge dinosaurs browsed and hunted within the forests and around their margins. Small mammals, some diurnal and some nocturnal, hunted the many insect types that flourished and foraged for seeds, living both in the undergrowth and in the trees. Some of the many species could also have lived semi-aquatic lives along the shallower water margins where horsetail stands grew, dwarf crocodiles hunted for small prey and seldom seen amphibians bred. On damp misty nights, when rain followed the hot dry seasons, the primitive frogs and salamanders emerged from their hiding places. By day large crocodiles basked by the deeper bodies of water, both fresh and brackish, in which turtles and fish with armoured enamel scales swam. Out along broad intertidal or seasonal lime-mudflats, a dazzling off-white in the intense sunlight, pterosaurs flew overhead and landed to filter-feed on small invertebrates living in shallow pools. Occasionally herbiverous dinosaurs crossed the flats *en route* for new feeding grounds, shadowed by their theropod predators.

A piece of the plastron or underside shell of a turtle with a hole possibly left by the tooth of a predator (left), part of a turtle pelvic girdle (top centre), a centrum of a dinosaur caudal or tail vertebra (bottom centre) and pieces of crocodile scute (right) from the Cherty Freshwater Member of the Purbeck Limestone Group, Durlston Bay, Dorset. Specimens: Author's collection. Photo: Author.

Such a reconstruction may well have its flaws, but there is evidence for each element of it. The evidence from Dorset stands on its own here, without need for recourse to the significant but less comprehensive material from Wiltshire – the sparser finds from the Vale of Wardour and Swindon indicate similar conditions without adding any significant details. This situation was to change with a remarkable discovery made in about 1980.

2
Wiltshire's Plant & Reptile Bed

South-west Wiltshire's Vale of Wardour is an arrowhead-shaped area of rolling hills and valleys sandwiched between chalk downlands to the north and south. The arrowhead points eastwards towards Wilton and Salisbury, its axis broadly followed by the River Nadder on its winding course towards its confluence with the River Avon at Salisbury. Many millions of years ago, as compression between the colliding African and European tectonic plates led to the formation of the Alpine mountain range, the Earth's crust underwent subsidiary folding along east-west axes in southern England. In the Vale of Wardour the River Nadder has exploited the weak and fractured rocks at the top of an upward fold or anticline to follow the course of least resistance. It is a classic example of a geographical feature known as inverted topography, and it has left exposed in the centre of the vale a succession of Upper Jurassic rocks deposited during the Kimmeridgian and Tithonian ages.

From the heart of the vale at Tisbury the River Nadder flows down past the hamlets of Upper and Lower Chicksgrove, through an area with a long history of quarrying of the local Portland Stone and, to a lesser but nonetheless significant extent, of Purbeck Limestone. Within Tisbury parish alone over forty stone quarries have been recorded. Research has suggested that stone from Chilmark, Tisbury, Tisbury Row and Upper Chicksgrove was used in the construction of Salisbury Cathedral, although Chilmark has traditionally claimed credit as the main source. The name Chilmark Stone is frequently used to describe high quality freestone from the main Portland building stone horizons. The industry died out briefly during the 1970s, but stone has since been extracted at Chilmark Ravine and Upper Chicksgrove.

Nineteenth-century workings at Upper Chicksgrove, long since disappeared in the wake of later quarrying operations, have an interesting place in the early history of English geology, and indeed of women in science. Although very much in a minority, women played a noteworthy role in the heady days of early palaeontological discovery in southern England. In Hampshire the gambler and traveller Barbara Rawdon Hastings, known formally as the Marchioness of Hastings and less formally as the 'jolly fast marchioness',[131] specialised in Tertiary fossils and assembled a collection that included some fine crocodile skulls. She was born in 1810, only a year before the less nobly born Mary Anning at Lyme Regis collected her first important marine reptile specimen, moving on to become one of the most significant collectors in the country. Wiltshire's offering, somewhere between the other two on the social scale but closer to the marchioness, was Miss Etheldred Benett (1776-1845) of Pyt House near Tisbury. 'Arguably the first female geologist',[132] and thought of in some circles as 'English Geology's First Woman',[133] Miss Benett 'did not succumb to the usual conventions of a woman in country society'.[134] Assembling an extensive collection of fossils was more to her taste, and in studying

---

131 Shindler (2010),

132 eNotes *Benett, Etheldred (1776-1845)* http://www.enotes.com/earth-science/benett-etheldred (visited 10 April 2010).

133 Winchester (2001), 117.

134 Cadbury (2001), 42.

the rocks at Upper Chicksgrove she 'made one of the very first bed-by-bed stratigraphical descriptions in the literature',[135] signed by her and now in the Geological Society of London Library. Miss Benett met William Smith, whose extensive fieldwork led to the publication of a coloured geological map of England and Wales in 1815, and presented him with an example of one of the better known fossils from the Vale of Wardour, the Tisbury Coral.[136] She contributed to an 1816 publication, *Sowerby's Mineral Conchology*, and in 1831 her *Organic Remains of the County of Wiltshire* was privately printed. Becoming 'so well known that eventually her name alone in the literature would suffice to denote work of outstanding quality',[137] Miss Benett was noted for her expertise on the fossil molluscs and sponges of the county.

The work of such early pioneers helped piece together a broad picture of the geology of southern England, within which the Jurassic and Cretaceous stratigraphy of the Vale of Wardour could be correlated with that of Dorset despite considerable local variation. When the geology of south Wiltshire was described in detail in 1903,[138] the basic succession of Kimmeridge Clay, Lower Portland Beds (Portland Sand), Upper Portland Beds (Portland Stone), Lower Purbeck Beds, Middle Purbeck Beds and Upper Purbeck Beds, all considered Upper Jurassic at that time despite some contention over the affinities of the Purbeck beds, was the same in Wiltshire as in Dorset despite considerable variations in thicknesses. Minor divisions, however, were described on a more local level. Four basic sub-units of the Portland Stone were recognised in the Vale of Wardour, although the classic locality for these was Chilmark Ravine rather than the Tisbury area: the Lower, Main or Chief Building Stones; the Ragstones; the Chalky Series; and

135 Benton, Hooker and Cook (2005), 54.
136 Winchester (2001), 114.
137 Cadbury (2001), 43.
138 Reid (1903), 4-29.

An ammonite from the Wockley Member of south-west Wiltshire's Portland Stone Formation. These large marine molluscs swam in the warm, shallow Upper Jurassic seas of what is now south-central England. The early nineteenth-century pioneering female geologist Miss Etheldred Benett was an expert on Wiltshire's fossil molluscs. Specimen: Author's collection. Photo: Author.

the Upper Building Stones. A Basement Bed at the bottom of this sequence, marking a possible fifth sub-unit, does not appear to have been recognised at Chilmark.

Pre-1976		Wimbledon 1976		Brit. Geol. Survey 1999	
Lower Purbeck Beds (Part of)		(N/A)	(N/A)	Purbeck Formation[1] (Part of)	Oakley Marl Member[2]
Upper Portland Beds or Portland Stone	Upper Building Stones	Portland Stone Formation	Chilmark Member	Portland Stone Formation	Chilmark Member
	Chalky Series		Wockley Member		Wockley Member
	Ragstones				Wockley[3] or Tisbury[4] Member
	Main or Lower Building Stones	Portland Sand Formation	Tisbury Member		Tisbury Member
	(Basement Bed)		Chicksgrove Member		
Lower Portland Beds or Portland Sand		Wardour Member		Wardour Formation	(Undivided)

[1]Now Purbeck Limestone Group   [2]Named by C.R. Bristow in 1995.   [3]Upper Chicksgrove   [4]Chilmark Ravine

Table 5: Portland and basal Purbeck succession in the Chilmark and Tisbury area.

In 1976 the stratigraphy of the Portland Beds of the Vale of Wardour underwent a thorough revision in a paper by W.A. Wimbledon, and 'previously quoted units, the 'Basement Bed', 'Main' and 'Upper Building Stones' and 'Chalky Series'" were set aside.[139] In keeping with the now accepted division of geological strata into 'formations' and subsidiary 'members', two formations with five members were described. The Lower Portland Beds became the Wardour Member of the Portland Sand Formation, and the little known and poorly described Basement Bed and the Lower Building Stones became the Chicksgrove Member and the Tisbury Member of the Portland Sand Formation. The movement of the major building stone horizons from a 'stone' to a 'sand' formation may at first appear

139 Wimbledon (1976), 3.

strange, but the glauconitic sandy limestones that make up these beds are very different from the famous white limestones of the Dorset Portland Stone, having a high sand content ranging up to 40 per cent or more and a honey-to-greenish colour imparted by particles of the mineral glauconite. Some of the beds have been referred to as calcareous sandstones rather than sandy limestones. The beds above the Tisbury Member, formerly known as the Ragstones and Chalky Series, became the Wockley Member of the Portland Stone Formation, overlain locally – at Chilmark Ravine but not at Upper Chicksgrove – by oolitic limestones of the former Upper Building Stones. These now became the Chilmark Member of the Portland Stone Formation. Thus did the Portland stratigraphy of the Vale of Wardour embark on the journey taken by the Purbeck stratigraphy of Dorset, i.e. into a state of some confusion.

Meanwhile, according to Wiltshire records, the quarry at Upper Chicksgrove had been notified as a site of special interest in 1971 under the terms of an Act of 1949. At this stage the quarry was significant on three counts: stratigraphically for its Portland-Purbeck exposure; palaeontologically for its ammonites, principally as 'the source of the type material for the original description of *Ammonites giganteus* Sowerby'[140] – an exceptionally large species of ammonite that now goes under the name *Titanites giganteus*; and historically for the work of Miss Etheldred Benett. In 1980 a 'plant and reptile bed' was discovered that would make it significant on a fourth count. This took the form of a fine sand or silt bed, deposited into a trough-shaped depression some 15m by 7m that had been eroded into the top of Wimbledon's Tisbury Member. Within this sand a large assemblage of plant, fish, reptile and mammal remains was discovered. The bed has been variously described in the limited literature, informally as 'the plant and reptile bed of the Wardour Portland Limestone' or

---

140 Wiltshire County Council records.

the 'plant bed'[141] and more formally as the Chicksgrove Plant Bed.[142] The author has also heard the term 'mammal bed' used informally in conversation with geologists in 2002. In the absence of a widely used formal name that adequately indicates its mixed floral and faunal contents, the term 'plant and reptile bed' is employed here. Evidence of plant and reptile remains at this horizon was not completely new, as Wimbledon had already described the basal bed of the Wockley Member at Upper Chicksgrove as 'a yellow based, grey bioclast sand' which 'contains abundant plant remains, and large reptilian bone fragments in the yellow base'.[143]

What was discovered in 1980 was by all known accounts, which at the time of writing in 2010 are very few, an exceptional assemblage of animal and plant fossils of predominantly lagoonal and terrestrial origin. A major excavation took place, assisted by dozens of volunteers 'from Salisbury, Southampton, Oxford and doubtlessly further afield'.[144] In 1987 the Upper Chicksgrove site was notified as a Site of Special Scientific Interest (SSSI) under the 1981 Wildlife and Countryside Act. The reasons for notification given to Wiltshire County Council included the following: 'The site has provided remains of 'dinosaur' genera, pterosaurs, crocodilians, sharks and bony fish as well as newly discovered multituberculate and pantothere mammals', and 'These vertebrates make this locality unique in the Jurassic succession of Europe.' The international significance of the site was made clear, as the rare mammal discoveries were described as being of an age somewhere between earlier Kimmeridgian mammal-yielding strata at Guimarota in Portugal and the Purbeck Mammal Bed of Dorset, and 'may equate

---

141   Benton, Hooker and Cook (2005), 55.
142   Astin (1987), 73.
143   Wimbledon (1976), 5.
144   Trevor Dykes *Mesozoic Eucynodonts* http://home.arcor.de/ktdykes/meseucaz.htm (visited 10 April 2010).

in age with the famous Morrison Formation of North America, with its dinosaur and mammal assemblages'.[145]

An assortment of reptile teeth and bones from the plant and reptile bed. An assemblage of plant, dinosaur, crocodile, pterosaur and mammal remains was first reported from this horizon in the 1980s, but full descriptions have yet to be published. Although fragmentary, the fossils provide evidence of a rich and diverse fauna. Specimens: Author's collection. Photo: Author.

In 1987 the first scientific paper on the plant and reptile bed was published, written by T.R. Astin and entitled *Petrology (including fluorescence microscopy) of cherts from the Portlandian of Wiltshire, UK – evidence of an episode of meteoric water circulation*.[146] It had been written as the result of 'a multidisciplinary study of the whole bed, including its provenance, stratigraphy and palaeontology', the overall study being referenced as 'Wimbledon *et al.*, in preparation' – in other words, this was to be just the beginning. Astin's specialised paper was

---

145 Wiltshire County Council records.
146 Astin (1987).

an investigation of chert found within the bed 'for information on the depositional and early diagenetic environments in which the original sediment was deposited'.[147] Diagenesis is physical or chemical change undergone by a sediment after deposition.

Some basic stratigraphic information was provided by Astin. The maximum thickness of the plant and reptile bed, referred to as the Chicksgrove Plant Bed, was given as 60cm, and it was illustrated in section as corresponding to the Chicksgrove 'Bed 25' of Wimbledon's 1976 paper. This was the basal Wockley Member bed containing lignite or carbonised plant remains. The overlying bed in the section, corresponding to Wimbledon's Chicksgrove 'Bed 26', was labelled 'Gastropod Micrite', i.e. a microcrystalline limestone with spiral shells, and referred to in the text as 'a lagoonal micrite'. It was an unevenly distributed bed which has been described since as 'a discontinuous gastropod-rich micrite'.[148] It is referred to here as the 'gastropod micrite'.

The significance of lagoonal deposits occurring within the Portland Stone Formation was considerable, especially when seen in conjunction with the erosion surface at the top of the Tisbury Member. A marine regression had apparently occurred in Wiltshire that was unrecorded in Dorset, and it marked the earliest evidence of what was to become a far more widespread regression in the following Purbeck and Lower Cretaceous Wealden times. Also remarked upon by Astin was the palaeosol in the basal beds of what is now the Purbeck Limestone Group, some 8m above the plant and reptile bed, which he specifically referred to as the 'Great Dirt Bed' through reference to Francis's 1984 paper on the Dorset Purbeck palaeoenvironment.[149]

Fossils preserved within the chert of the plant and reptile bed were briefly described, and included gastropods, plant spores and

---

147 Astin (1987), 73.
148 Benton, Hooker and Cook (2005), 55.
149 Astin (1987), 73.

wood fragments. The wood lacked bark and showed evidence of having rotted in places before the silicification that took place during the early stages of chert formation. In other places some cell structure was preserved: 'The cellulose cell walls of the woods were replaced first ... and the cells subsequently infilled with brightly fluorescent silica.'[150] Fossils are often highly compressed before preservation, and plants are especially susceptible to this. It was noted of the fossils in the chert that conditions had 'led locally to the pre-compaction preservation of fossils with obvious palaeontological benefits'.[151]

If the detailed chemistry and mineralogy of Astin's paper is beyond the level of this work, the postulated sequence of events by which chert containing fossil plant remains was formed is not. Following the return of marine conditions the plant and reptile bed and the gastropod micrite were buried by overlying marine limestones, and the pore spaces in these deposits accumulating beneath the sea bed were filled with seawater. When the sea receded and the dry land which led to formation of the Great Dirt Bed or its Wiltshire equivalent emerged, rainwater penetrated the surface to form groundwater. This at first mixed with and then replaced the saltwater as it permeated down through the beds. The sand of the plant and reptile bed was less permeable than the limestone deposits and 'likely to have caused a change to lateral flow of the meteoric water'.[152] These specific conditions favoured the dissolution of carbonates in the bed and a three stage process of silicification began via opal, length-fast chalcedony and quartz formation. The final stage of the process occurred either when 'marine modified' conditions returned and 'vigorous groundwater circulation ceased' or at a later time.[153] The 'marine modified' conditions presumably refer to the variably saline

---

150   Astin (1987), 80.
151   Astin (1987), 84.
152   Astin (1987), 82.
153   Astin (1987), 82.

lagoonal or 'inland sea' environments under which the overlying Purbeck limestone beds were deposited.

Although of considerable interest to sedimentologists, Astin's paper gave nothing away about the exciting fauna outlined in the 1987 SSSI notification. Dinosaur, pterosaur and mammal finds from an English Tithonian Age deposit were of exceptional palaeontological significance, and publication of the multi-disciplinary paper 'in preparation' was thus to be awaited with eager anticipation. A book on vertebrate fossils published in 1995 included a chapter on reptiles from the Portland Stone of England co-written by Wimbledon,[154] and at last some long awaited answers could be expected.

References were made to a fauna 'recently discovered in a mid-Portlandian stratum in the Vale of Wardour' that included remains of *Goniopholis*, dwarf crocodilians such as *Bernissartia*, pterosaur bones and teeth, *Nuthetes* and lepidosaurians.[155] It is reasonable to deduce that the plant and reptile bed was the stratum in question. No turtles had apparently been recorded from the Portland Stone of Wiltshire, and dinosaur remains included 'dorsal spines of an unidentified armoured dinosaur from the "higher" Portlandian beds at Swindon'.[156] A table of reptile distribution within the Portland Beds included no chelonians (turtles), pterosaurs, sauropods, theropods (despite the reference to *Nuthetes*) or ornithopods from Wiltshire,[157] although reference was made in a separate table of Wiltshire Portland Bed reptiles to 'saurian' vertebral centra from Chicksgrove and Pyt House dating back to the days of Etheldred Benett.[158] All in all, the information provided in the chapter was disappointingly sparse.

154 Delair and Wimbledon (1995).
155 Delair and Wimbledon (1995), 354.
156 Delair and Wimbledon (1995), 354.
157 Delair and Wimbledon (1995), 355.
158 Delair and Wimbledon (1995), 352–3.

An exceptionally well-preserved shoot, probably of an extinct conifer, embedded in silt of the plant and reptile bed. The bend in the stem indicates that the wood tissue was still relatively soft at the time of preservation. The shoot was probably silicified when a flow of silica-bearing underground water slowed with a change in direction.
Specimen: Author's collection. Photo: Author.

As the rest of the 1990s passed by no further revelations about the plant and reptile bed appeared, although the stratigraphy underwent some revisions. In 1999 a memoir was published to accompany the British Geological Survey's 1:50,000 scale map of the Wincanton district. When mapping the geology of the district eastwards to Tisbury, through which the boundary with the Salisbury district passes, the British Geological Survey worked to an amended version of Wimbledon's stratigraphy (see Table 4). The Wardour Member of the Portland Sand Formation had become a localised formation in its own right, the Wardour Formation. All overlying strata to the approximate base of the Purbeck were incorporated into an expanded Portland Stone Formation, thus restoring the original Upper Portland Beds or Portland Stone of south-west Wiltshire as a single stratigraphical unit – a unit dating back at least to 1896. Beds to the west of Tisbury identified as 'possibly the Chicksgrove Member of Wimbledon'[159] were now included in the Tisbury Member. Where mapping problems arose, the lower beds of Wimbledon's Wockley Member corresponding to the old Ragstones were also incorporated into the Tisbury Member. Wimbledon's junction had proved to be 'readily recognisable in quarry sections, but not always easily identifiable from brash'.[160] Brash is the weathered broken rock near the surface on which mappers usually depend in the absence of quarry or cliff exposures.

What remained of Wimbledon's Wockley Member in the British Geological Survey's Wincanton district stratigraphy largely corresponded to the former Chalky Series of the Portland Stone, and the lower part of what is now the Purbeck Limestone Group was referred to as the Oakley Marl Member of the Purbeck Formation. Fortunately, through all this juggling around, the junction between the Tisbury and Wockley Members at Upper Chicksgrove appears on balance to have remained unaltered, but the problems of mapping an area of locally variable near-shore sediments were clear to see.

In 2001 *Mesozoic and Tertiary Palaeobotany of Great Britain* highlighted the dearth of surviving Upper Jurassic palaeobotanical exposures in southern England, stating the only significant ones to be 'the 'fossil forest' localities in the dirt beds of the Purbeck Limestone Formation'[161] and pinpointing the Isle of Portland as the best place for their study. Although reference was made to the Portesham Quarry flora, it was not deemed worthy of inclusion as a Geological Conservation Review site.

In 2004 the Portland Stone dinosaurs of Wiltshire were briefly listed in the second edition of the major work *The Dinosauria*, with no specific reference to locality. Included were indeterminate or undescribed armoured dinosaur remains of the Stegosauria

---

159 British Geological Survey (1999), 65.
160 British Geological Survey (1999), 68.
161 Cleal, Thomas and Batten (2001), 103.

and Nodosauridae, indeterminate Ornithischia, *Echinodon* sp. (the heterodontosaurid ornithischian originally described from the Purbeck Limestone Group of Dorset), 'Theropoda indet. (=*Megalosaurus* sp.)', and 'Sauropoda indet. (=*Camarasaurus* sp., *Diplodocus* sp.)'.[162] Sauropods were appearing on the scene, indeterminate but associated with specific genus names that brought more to mind the dinosaur graveyards of the American west than the rolling countryside of south-central England.

In 2005, with still no sign of any further scientific papers on the plant and reptile bed appearing, a sister volume to the 2001 palaeobotany book entitled *Mesozoic and Tertiary Fossil Mammals and Birds of Great Britain* was published.[163] Ironically this gave at least some information on the plant and reptile bed flora, which had not received a mention in the plant volume. Drawing on an anonymous unpublished 1983 report to the Nature Conservancy Council,[164] the relevant chapter in the volume referred to 'abundant remains of current-aligned flattened carbonized plant material', to silicified fossils including 'seeds, wood and gastropods' and to 'megaspores and microspores, conifer seeds and wood'.[165] If the flora was beginning to look as if it could be of some considerable significance, however, the spotlight would inevitably focus on the references to the reptiles and mammals.

Information on the reptile assemblage was more detailed than that provided in the 1987 Wiltshire County Council SSSI notification, and went beyond the limited 1995 references and even the 2004 dinosaur list. It was now said to include 'bones and teeth of ... crocodilians, pterosaurs, ornithischian and saurischian dinosaurs, lepidosaurs ...'. The lepidosaurs or lizards were represented by the

162   Weishampel et al. (2004), 546.
163   Benton, Cook and Hooker (2005).
164   Anon (1983).
165   Benton, Hooker and Cook (2005), 55.

smallest of the reptile teeth.¹⁶⁶ Three types of crocodile were present in the form of goniopholids, theriosuchians and *Bernissartia*, two pterosaur species were assigned to the genera *Pterodactylus* and *Gnathosaurus*, and dinosaurs identified from teeth included 'one diplodocid sauropod, a camarasaurid sauropod, small and large theropods, presumed coelurosaurs, a fabrosaurid and *Iguanodon*'. Armoured forms, either nodosaurs or stegosaurs, were also found. A remarkable picture of life on land during Tithonian times in southern England could now be drawn, even on the very limited information released. Where exactly the land was and how much of it there was in south Wiltshire at any given time during the formation of the Portland Stone is not known, as much or all of the material had been transported by water from elsewhere before burial – features of the silicified wood studied by Astin were interpreted as having resulted from 'transport and ... early burial'.¹⁶⁷

Most elements of the diverse reptile fauna will be looked at more closely in later chapters, but three with a key place in the early

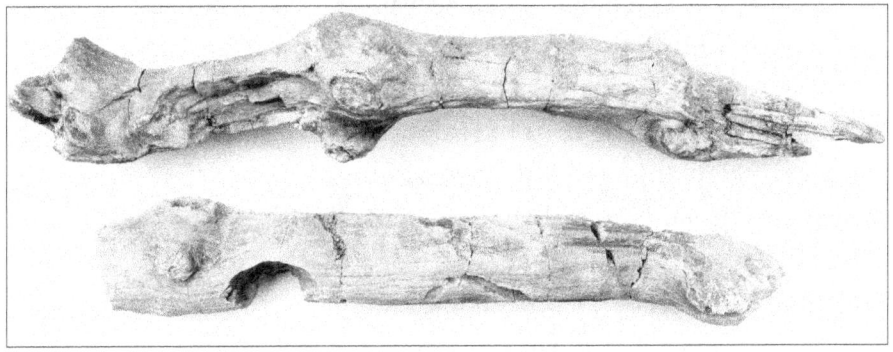

Two unusual lengths of silicified wood preserved in the base of the plant and reptile bed. Nodes with the truncated bases of shoots and small branches provide evidence of a distinctive growth pattern. Even such relatively robust specimens require careful collecting and reassembly if damage is to be avoided. Specimens: Author's collection. Photo: Author.

166 Benton, Hooker and Cook (2005), 55.
167 Astin (1987), 79.

English and European history of the discovery of the Age of Reptiles are worthy of a brief mention here. *Bernissartia* – a genus to which, as mentioned in the preceding chapter, some button-like teeth from the Purbeck Limestone Group of Dorset possibly belong – was named after Bernissart in Belgium, where it was discovered in 1878 in Early Cretaceous strata in a coal mine. The coal mine is of greater significance for the discovery at the same time of numerous *Iguanodon* skeletons, which rank among the earliest finds of complete dinosaur skeletons. The beaked herbivorous *Iguanodon* was only the second dinosaur to be described, by Mantell in 1825 on the basis of fragmentary material from southern England. *Pterodactylus* was the first flying reptile to be identified as such, named by French naturalist Georges Cuvier in 1809. Represented by a range of short-tailed species, the first English specimen of the genus was discovered by Mary Anning in 1828.[168]

It was the mammal finds that led to publication of the report on the plant and reptile bed in the 2005 mammal and bird volume. Their occurrence at Upper Chicksgrove had led to the official designation of the quarry as a Geological Conservation Review or GCR mammal site, and yet information on them remained almost as obscure as ever. Coming on for a quarter of a century after their discovery they had 'not been studied yet, so their full import has to be determined', despite representing an apparently diverse assemblage from the only Upper Jurassic mammal site in Britain. The only snippet of information of substance was that members of three orders had been positively identified: Multituberculata, Triconodonta and 'Eupantotheria'.[169] Suffice it to say for the purposes of this description that a number of probable seed-eating, insectivorous and carnivorous small mammals lived in the terrestrial environments from which the plant and reptile bed material was at least partly derived.

---

168 NHM www.nhm.ac.uk/nature-online/science-of-natural-history/biographies/mary-anning/index.html (visited 16 May 2010).
169 Benton, Hooker and Cook (2005), 54–6.

The international significance of these finds was once again emphasised in the 2005 publication, where they were now described as filling a gap of approximately 3 million years between older mammal and reptile faunas from Guimarota in Portugal, the Morrison Formation of the USA and Tendaguru in Tanzania and the younger faunas of the Lower Cretaceous beds of the Purbeck Limestone Group of Dorset.[170] With reference to the mammals, it has been stated of the Upper Chicksgrove site: '...it's definitely the most significant source on the planet. This Tithonian fauna's a world leader as there isn't another known competitor anywhere on Earth.' The comment comes from Trevor Dykes's 'Mesozoic Eucynodont' website, the website 'where ancient mammals defy the dinosaurs', and it seems unlikely he is the only one mystified at the failure to publish proper descriptions after such a length of time. Modern palaeontology is a fast-moving international discipline, and those who follow it with interest are not immortal. Referring to the dozens of volunteers who helped at Upper Chicksgrove in the early 1980s, Trevor Dykes pointedly remarks that 'It would be kind of nice if that assistance could be appropriately respected and rewarded, with proper identifications and descriptions of at least some collected fossils, and before all of those volunteers have had time to drop dead and fossilize for themselves.'[171] As it was, the UK government-run JNCC (Joint Nature Conservation Committee) felt obliged to designate Upper Chicksgrove Quarry a GCR site for Jurassic mammals on the basis of an unpublished and anonymous report, backed by an assertion that the finds had not yet been studied – although they had apparently been studied in sufficient depth some eighteen years previously to enable the reporting of 'newly discovered multituberculate and pantothere mammals' in the SSSI notification (see this chapter above).

170 Benton, Hooker and Cook, (2005), 54–6.
171 Trevor Dykes *Mesozoic Eucynodonts* http://home.arcor.de/ktdykes/meseucaz.htm (visited 10 April 2010).

Forests of the Dinosaurs

A tradition of amateur fossil collecting in England, often by individuals who contribute to scientific descriptions and support the work of professionals, has played a significant role in palaeontology since the days of Mantell. The fascination of fossil collecting can make it a compulsive habit that, as has already been seen, has been no respecter of sex or class since the early nineteenth century. For this author it was very much an abandoned habit by 2002, bar the occasional opportunist collecting for old times' sake, and had been so for over a quarter of a century since the days when he collected shark teeth from Tertiary deposits in the south-east of England. Geological interest had meanwhile shifted to natural stone as a masonry product, and the building stones of the Vale of Wardour were under study as the subject of a planned but as yet uncommissioned book. It was background research for this project that led by chance to a dramatic revival of the interest in fossils. During a visit to a local exposure of Portland building stone strata, the discovery of a bed of silty sand with strange looking off-white fossils projecting from it, mistaken at first for small bones, marked the moment of truth. With the collection of a small bag of silt for further examination at home, the gateway to a world of

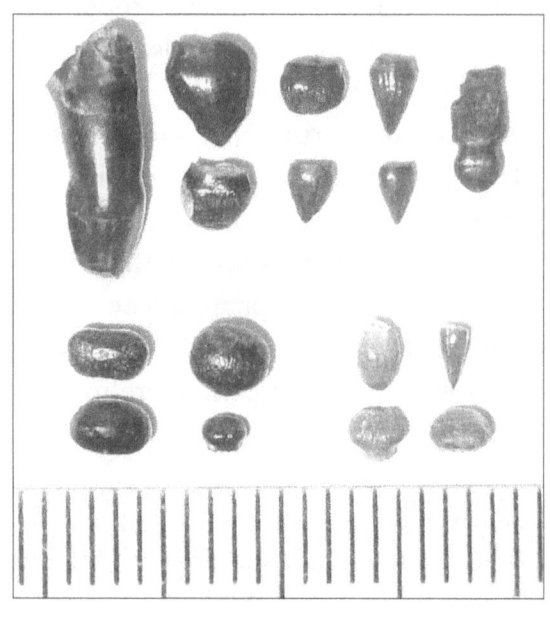

Eight teeth typical of the small heterodont atoposaurid crocodile Theriosuchus (top) and four bun-like crushing teeth typical of the small crocodile Bernissartia (bottom left) from the Purbeck Limestone Group of Durlston Bay, compared with two teeth of cf. Theriosuchus (bottom right, upper) and two teeth of cf. Bernissartia (bottom right, lower) from the plant and reptile bed. Specimens: Author's collection. Photo: Author.

discovery was opened. The projected book on the building stones of the Vale of Wardour was side-tracked into obscurity.

With a reasonable knowledge of the local stratigraphy already obtained through study of the building stones, and with an examination of Wimbledon's 1976 paper in Salisbury library, it was not difficult to identify the horizon from which the fossiliferous silt had been collected as that of the plant and reptile bed. Meanwhile an attempt was made to retrieve its fossil contents. There seemed to be little to do with it but to sieve it, which left a residue of mostly broken fragments and numerous minute gastropods. It rapidly became clear that the fragments were not bone but petrified plant remains. A finely preserved seed clinched the matter, but apart from this and a couple more seeds it was all but impossible to reassemble complete specimens of other plant parts. Occasionally a reasonable specimen could be put together, but always with one or more parts missing. A search through the smallest remnants under a magnifying glass revealed a few teeth typical of the extinct fish genus *Lepidotes*. These distinctive peg-like teeth are abundant in the Purbeck Limestone Group, and were used by their mollusc-eating owners to crush the shells of their prey. Also found was a single small crocodile tooth.

Over the next few years a number of opportunities to collect from this bed arose. Further finds and the acquisition of various published material – the 1987 Upper Chicksgrove Quarry SSSI notification, Astin's 1987 paper and the 2005 mammal and bird volume – made it possible to build up a much clearer picture of the bed. The material matched that described by Wimbledon in 1976 and the anonymous report of 1983 in terms of stratigraphy, lithology and many aspects of the palaeontology, but reached nothing like the thicknesses of up to 60cm previously recorded and was of very limited extent. The silty sand in fact did not normally exceed a maximum thickness of around 20cm. Most of this contained a large quantity of flattened moulds left by decayed plant material, with no more than a layer of loose grainy carbon particles remaining, and the areas with petrified material were confined to small lenses making up only a minority of the bed. A quantity of the silt without silicified fossils was bagged up for later processing for vertebrate remains.

The original plant and reptile bed exposed in 1980 had, according to a photograph of volunteers at work,[172] been collected from a well-levelled area. During eight years of intermittent collecting from later minor exposures of this bed, access was never easy as it was only exposed between large blocks of Tisbury Member limestone and overlying blocks of gastropod micrite. This compromised the ability to collect specimens in any great quantity and to do so without damaging them. Even under these circumstances, however, patience and the development of improved collecting techniques led to the assemblage of a remarkable collection of fossils, many of which had never been previously reported from the UK.

With sieving abandoned as a method of processing the silt, the only other option was to carefully remove a small area of gastropod micrite, to work slowly around an area of the underlying silt, and to prise it out in as few pieces as possible, carefully placing them side by side on a tray to make reconstruction of specimens split between two pieces more manageable. PVA glue was used for consolidating specimens on site: as a water-soluble material it was a reversible process that allowed for later cleaning. Back at home work continued with the aid of a magnifying table lamp, a magnifying glass and an assortment of small brushes, toothpicks and small plastic spatulas. PVA was used to glue the parts of fragmented specimens together, and also to consolidate silt around specimens that were left embedded in the matrix when seen as too delicate to risk attempting complete removal. Just how many hours went into all this was not recorded, but the results were more than sufficient reward.

---

172 Benton, Hooker and Cook (2005), 55.

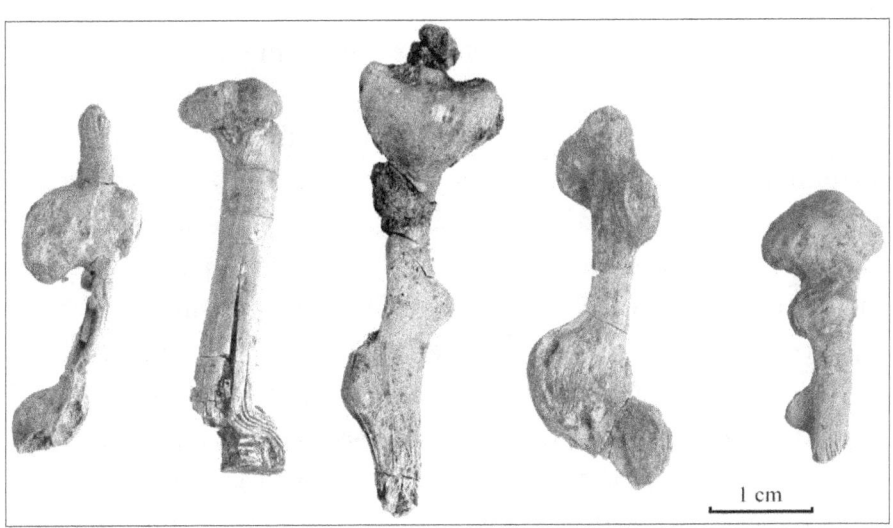

Delicate plant shoots from the plant and reptile bed removed in several pieces and reassembled using PVA glue. Removal of such material from the silt was carried out under an illuminated magnifying lamp using a fine brush, and specimens deemed too delicate were left embedded in PVA-consolidated matrix. Specimens: Author's collection. Photo: Steve Clifford.

Chiselling away at the gastropod micrite had meanwhile provided plenty of opportunity for a close-up study of the bed. Some sections came away without too much of a struggle, especially where fragmented around fissures, but larger blocks put up greater resistance unless they had been exposed to severe frosts. Worst of all were sections where the blocks were welded together by hard calcite deposits left by water percolating down from the overlying limestone beds.

Many of the fossil types found in the plant and reptile bed also occurred in the gastropod micrite, including, as would be expected, some of the small gastropods. Empty moulds of plant fragments were common in the base of the micrite, decreasing in frequency towards the upper parts of the bed, and occasional indeterminate reptile bones and teeth of crocodiles and theropods were also to be found in the lower part of the micrite. Other than small fragments, the remains of large bones were in fact *only* observed in the micrite and not in the plant bed itself, and then only on a very localised basis. These remains were generally very soft and extremely difficult to extract from the surrounding stone. A concentration of such material directly over a lens of the plant bed suggested that the plant and reptile bed and the base at least of the gastropod micrite resulted from a single depositional event, and the manner in which the beds graded irregularly into each other supported this view. In the original excavations the plant and reptile bed had been described as laminated,[173] and personal observation confirmed this to be true of most of the newly exposed section, especially at the base where some medium-sized pieces of fossil wood aligned on an approximate north-south axis were found. However, there were also patches and mounds where current alignment was along different axes or non-existent, with some plant remains even deposited vertically in non-laminated sand. The appearance here was of reworked material or of deposition from highly localised turbulence spreading outwards to layered deposition in calmer surrounding water. Such distribution, i.e. 'the presence of randomly oriented organic matter in the sediment',[174] is one of the recognised criteria for single depositional events. The possibility of bioturbation, or that the sediment had been disturbed by animals, can also be considered. The three-dimensional preservation of the petrified plant fossils physically reflects the different positions in the sediment, ranging from strong lateral compression in current-aligned and horizontally laid specimens to less obvious vertical compression in vertically laid specimens.

Persistent collecting that focused for the most part on plant remains was also rewarded with some significant reptile finds between 2008 and 2010. The first specimen of note to be obtained was an incomplete sauropod tooth crown from the gastropod

---

173 Benton, Hooker and Cook (2005), 55.
174 Martin (1999), 230.

micrite, to be followed soon after by a sizeable crocodile tooth from the plant and reptile bed itself. The sauropod tooth was submitted to the Natural History Museum in London for identification, and returned labelled 'cf. *Pelorosaurus*'. This was a comparison rather than a definitive identification, and despite having been reported from the Purbeck Limestone Group of Buckinghamshire, *Pelorosaurus* is often considered to be a Lower Cretaceous sauropod – according to the caption accompanying tooth images available from the NHM, it 'lived 130 to 112 million years ago'.[175] The teeth were, however, indicative of possible brachiosaurid dinosaurs inhabiting Tithonian Wiltshire.

An increased range of reptile teeth and small bones in varying states of preservation was to follow in 2010. Most notable was a complete sauropod tooth crown with excellent surface detail. The crown was spoon-shaped and the enamel bore several scratch marks. The specimen included a delicate collar of root material at the base, most of which was salvaged through careful collection and consolidation but would have been lost with sieving. A small de-enamelled crown of a tooth with similar basic morphology, small theropod teeth, pieces of broken bone, a pterosaur tooth and a pterosaur bone, probably the ultimate phalange of the fourth or wing finger, were also collected along with turtle fragments, several crocodile teeth and a range of as yet unidentified material. The turtle fragments were unexpected, having been reported as absent from the Wiltshire Portland Beds in 1995.

Some reptile remains from the upper beds of the underlying Tisbury Member were also collected between 2002 and 2010. Specimens obtained by the author include a pterosaur bone and part of a large reptile bone preserved in chert. Although otherwise indeterminate, the pterosaur bone is a possible phalanx from an eagle- to albatross-sized flying reptile. The precise bed from which these fossils originated is unknown.

175 NHM http://nhm.ac.uk/piclib/www/image.php?img=51144&frm=ser&search =teeth (visited 16 May 2010).

A pterosaur bone from the Tisbury Member of south-west Wiltshire's Portland Stone Formation. It is possibly a wing phalanx of a flying reptile the size of an eagle or albatross, but is otherwise not determinate. The surrounding stone is considerably harder than the delicate bird-like bone. Identification: NHM. Specimen: Author's collection. Photo: Author.

By the summer of 2010, to the author's knowledge, no further collecting from the plant and reptile bed was practical from known exposures. With regard to the plants, what had been collected since 2002 included several previously undescribed seed and shoot species and a range of other as yet unidentified material. Also collected were specimens that could be attributed to previously described morphotaxa from other localities, as will be seen in the following two chapters. The vertebrate fossils were a valuable if limited taster of the wider assemblage collected in the early 1980s and tantalisingly described in outline in 2005, even if only on the evidence of an unpublished and anonymous report. In contrast to the Purbeck Fossil Forest with its autochthonous or *in situ* fossils, the plant bed had the disadvantage of only containing allochthonous or transported material, and hence the fragmentary nature of the remains. Its advantage was that as a detritus bed a large amount of material had been sorted and concentrated through the action of moving water, leaving a more taxonomically diverse range of fossils. The inevitable question arose as to whether the fossil contents of the plant and reptile bed and the Purbeck Fossil Forest complemented each other in building a picture of the Tithonian environment of southern England, or whether they originated from quite different environments.

3
THE DORSET CONNECTION

As the science of geology has become ever more complex it has spawned numerous specialist areas of research. Palaeontology was one of the earlier branches, incorporating the study of fossils of all types. A later offspring of palaeontology was taphonomy, or the science of fossil preservation, which deals with the pathways organisms follow from death to fossilisation. In the case of plants the end result of these pathways ranges from carbonisation, including charcoal preservation, to petrifaction. Petrifaction or petrification means conversion into stone, but in palaeobotany petrifaction can be used with specific reference to 'the addition of mineral matter to existing hardparts, especially porous ones such as ... wood'. Where existing parts are *replaced* by mineral matter the word replacement can be used, and permineralisation refers to 'fossilization by the precipitation of dissolved minerals in the interstices of hard tissue'.[176] Nonetheless, as 'the term petrifaction has sometimes been used synonymously with permineralization',[177] and as petrification is synonymous with petrifaction, the terms can lead to some confusion. All the plant fossils mentioned in this chapter can safely be referred to as petrifactions, and of the various forms of petrifaction, silicification

---

176 Brown (ed.) (1993), 2166.
177 Martin (1999), 133–4.

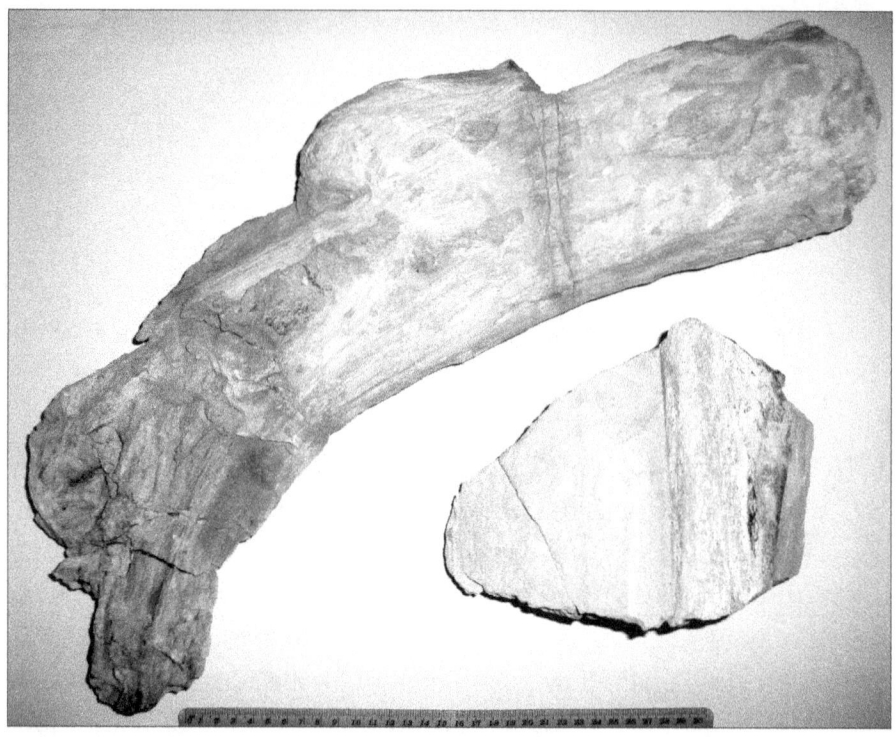

Conifer wood petrifaction from the plant and reptile bed (above) and reptile bone petrifaction embedded in chert from the Tisbury Member of the Portland Stone Formation (below). The process of silicification requires a source of readily soluble silica, which is deposited into the bone or wood from solution in ground water. Specimens: Author's collection. Photo: Author.

is the name for the process when the mineral involved is silica (SiO_2), calcification when the mineral is calcium carbonate ($CaCO_3$), and pyritisation when the mineral is iron pyrite (FeS_2).

A beautifully preserved fossil can belie the often less than pleasant processes through which it reached that state. To one expert, describing the way in which the science of taphonomy is viewed by those with only the vaguest understanding of it, 'it deals with death, decay, and disintegration, and is the science of dead, rotting things accompanied by a terrific stench'.[178] This romanticised view could well apply to the contents of the fish bed at the base of the Purbeck Limestone Group in south-west Wiltshire, where the density of specimens indicates a possible 'sudden death assemblage' resulting from one of a number of causes, such as a change in salinity or the drying out of isolated pools. In the plant and reptile bed, however, most of the material is debris in the form of detached plant remnants, scattered bone fragments and teeth that would largely have been shed during life.

An examination of silicified plant fossils from the plant and reptile bed does not bring to mind images of decaying vegetation, because several of the specimens preserve the most extraordinary surface detail and delicacy of features. This suggests that they were buried quickly before the processes of decay had set in, and consequently that the many other fossil plant remains that do show signs of wear and decay had reached that condition before arrival and subsequent deposition in the bed.

In geological deposits laid down over long periods of time on the beds of seas and lakes the fossils tend to be graded, a good example being seeds of the water plant *Stratiotes* found in approximately 35 million year old deposits on the Isle of Wight. At one locality the carbonised seeds show excellent preservation, and at another they are severely degraded and show signs of decomposition before fossilisation.[179] Each location shows evidence of one distinct set of conditions. Some deposits, however, such as point bar deposits made up of material laid down on the inside of river bends, are usually of short duration and eroded away before entering the geological record. In such deposits 'fossil plant remains may represent more than one depositional event and mixing of plants from different communities,

---

[178] Martin (1999), xiii.
[179] Collinson and Cleal (2001b), 253.

as suggested by age mixing ...'.[180] While not necessarily this type of deposit, the plant and reptile bed appears to have captured such a mix from different communities, incorporating fresh material and transported or redeposited material from indeterminate sources of origin. In studying the fossils from such a deposit, comparative material from other sites is required if any attempt to understand the environments in which they lived is to be made.

Such 'other sites' are not easy to come by, as Mesozoic strata yielding seed and shoot petrifactions are not abundant in the UK. Before looking for them, it is worth taking a brief look at an English Tertiary site where the study of fossil seeds was first given serious attention – a site that has long been famous in the annals of palaeobotany, and one that deserves a mention in any book on three-dimensionally preserved British fossil plants. This is the line of cliffs along the north coast of the Isle of Sheppey in Kent, 'one of the classic palaeobotanical sites in the world'.[181] Here, in deposits of London Clay laid down some 55mya during the Eocene epoch of the Palaeogene period, is to be found an extensive flora of fruits, seeds and wood largely preserved as pyrite petrifactions. They are mostly angiosperm remains from long after the extinction of the dinosaurs, and comparisons with Jurassic floras are thus of limited use, although some recorded finds such as that of a leafy *Araucarites* sp. shoot[182] would be quite at home in a Jurassic plant bed. It is of note that the number of seed species from the Isle of Sheppey, largely representative of existing plant families and genera, totals well over 300, as opposed to about ten species from the plant and reptile bed. This is the result in part at least of a long-term increase in plant diversity, most notably of seed-bearing plants, that has accelerated over the past 100 million years. Of the angiosperm species from Sheppey, 289 were described

---

180 Martin (1999), 230.
181 Collinson and Cleal (2001a), 191.
182 Collinson and Cleal (2001a), 195 Fig. 8.11.

The study of fossil seeds and fruits, such as these specimens from Wiltshire's plant and reptile bed, is known as palaeocarpology. This branch of palaeontology developed in the wake of a detailed description of Tertiary seeds and fruits from the Isle of Sheppey in Kent published in 1933, but few Jurassic seeds can be used in reliable whole plant reconstructions. Specimens: Author's collection. Photo: Author.

by Eleanor M. Reid and Marjorie E.J. Chandler in 1933 in a classic work entitled *The London Clay Flora*, which led the way in developing a further branch on the tree of geological sciences, in the form of palaeocarpology or the study of fossil fruits and seeds.

The impressive species diversity of the Isle of Sheppey flora allows a reasonable reconstruction to be made of an Eocene forest environment. Despite the passage of 55 million years, seeds can be attributed to families representing palm, mangrove, frankincense, tea, grape, walnut and magnolia plants, to name but a few. It is only in terms of the exceptional three-dimensional preservation that the flora is comparable to the plant and reptile bed seeds, and anyone who has seen a collection of seeds from the Isle of Sheppey – or from one of the other London Clay sites in south-east England – and a collection of seeds from the plant and reptile bed cannot fail to be impressed by the detailed preservation. It is almost as if a packet of mixed seeds could be put together ready for sowing, to create either an Eocene or Jurassic botanical equivalent of *Jurassic Park*. Unfortunately, although the most extraordinary detail can be preserved in plant fossils – an exceptional Jurassic specimen from India has even 'revealed chromosomes at different stages' in a dividing cell,[183] – beautifully preserved plant structures are 'not so well preserved that "defossilization" will ever be successful!'[184]

Despite the disappointing news that fossil plants will never be brought back to life, it is remarkable enough that plant seeds are very occasionally preserved in such fine detail. There is, to the author's knowledge, no comparable assemblage of three-dimensionally preserved fossil seeds to be found in Britain from the rocks that span the approximately 90 million years of geological time between the Tithonian deposits of the plant and reptile bed and Portesham Quarry and the Eocene deposits of the London Clay. In terms of plant extinctions and macroevolution, disproportionately far more happened during that 90 million years than during the last 55 million years, and it is for this reason that many Jurassic seeds are unattributable to known plant families or orders, living or extinct.

Exclude the plant and reptile bed material and the Purbeck Fossil Forest and the British Upper Jurassic plant record is very

---

183 Agashe (1997), 53–4.
184 Martin (1999), 133, with ref. to Jones, D. (1998) Defossilization. *Nature* 394: 727.

limited. In 2008 a fourteen-page overview of the UK's Jurassic fossil plant finds devoted less than 200 words to the Upper Jurassic of England, merely making mention of the conifers and bennettites of the latter.[185] The Geological Conservation Review (GCR) volume *Mesozoic and Tertiary Palaeobotany of Great Britain*[186] chose a single Purbeck Fossil Forest site on the Isle of Portland yielding only conifer remains to represent English Upper Jurassic palaeobotany. The GCR series, it should be noted, is an ambitious project initiated in 1977 as a 'systematic selection and documentation of a country's best Earth science sites'.[187] By its very nature it focuses on pinpointing a network of geological sites chosen as worthy for conservation, and has no remit to cover those that have been irretrievably lost or are unlikely to yield any further worthwhile fossils. It thus works under limitations, but even allowing for this the authors gave no hint of having being made aware of previous plant and reptile bed finds, or of any other similar English finds.

Looking at the UK as a whole, Jurassic plant sites are of limited geographical distribution and for convenience can be grouped into three areas, namely Yorkshire, southern England and Scotland. Fossils from the significant Middle Jurassic sites of Yorkshire's Cleveland Basin, 'one of the classic areas in the world for Mesozoic palaeobotany',[188] bear no apparent similarity to the plant and reptile bed specimens, and neither do the Middle and Upper Jurassic fossils from the sites in north-east Scotland and the Isle of Skye. The mode of preservation plays a key role. For the most part these fossils are compressions, largely of leaves, which at the best of times are difficult to correlate with petrifactions, and even where three-dimensional preservation does occur the specimens and modes of preservation

---

185 Van Konijnenburg-van Cittert (2008), 70–1.
186 Cleal, Thomas, Batten and Collinson (2001).
187 Ellis, N. V. in Cleal, Thomas, Batten and Collinson (2001), xvii.
188 Thomas and Batten (2001a), 31.

A fossil conifer wood specimen from the plant and reptile bed (left) invites comparison with a morphologically similar conifer wood specimen from a present day tree (right), the mountain pine Pinus mugo subsp. uncinata. Such apparently timeless structures can belie the changes that have taken place during the past 146 million years. Specimen: Author's collection. Photos: Steve Clifford, Author.

bear little obvious comparison to plant and reptile bed material. As for southern England, apart from the Purbeck Fossil Forest there are a grand total of two GCR sites, both Middle Jurassic and both yielding impressions highlighted by iron staining.[189]

The GCR plant volume further revealed the limited availability of Upper Jurassic plant fossils in continental Europe: 'Late Jurassic floras are generally very rare in Europe and none appears to be Portlandian in age.'[190] 'Portlandian', it should be noted, is now internationally redundant as the name for a geological age, having been incorporated into the Tithonian, but the conclusion from this is clear: the plant and reptile bed contains the only known available European flora from a substantial part of the Tithonian Age.

Two Scottish Jurassic GCR sites yield Middle and Upper Jurassic plant fossils variously described as 'permineralizations', 'petrifactions' and 'petrified plants'.[191] They are preserved in calcite nodules and retain fine detail that can be studied through sectioning and a process

189 Cleal, Thomas and Batten (2001), 104–8.
190 Cleal, Thomas and Batten, (2001), 112–13.
191 Thomas and Batten (2001c), 128–33.

known as cellulose acetate peeling. It is this process used on 'coal ball' carbonate nodules from the Carboniferous Age that led to a remarkably detailed understanding of some palaeophytic floras, but such permineralised Jurassic floras 'are globally rare'.[192] The fossils are encased in the nodules and three-dimensional reconstruction requires a lengthy sequence of peelings to be taken through the specimens, whereas in the plant and reptile bed fossils internal preservation is accompanied by preservation of the external three-dimensional surfaces.

The author obtained his first knowledge of Jurassic palaeobotany through the GCR plant volume. Still unable to identify any of the plant and reptile bed specimens, he followed up the reference to the Portesham Quarry paper and obtained a copy through Wiltshire inter-library lending. This was the opening to much of the further work that led to the writing of this book. It has already been related in the opening chapter that plant petrifactions were described by Brown and Bugg in the land plant section of the paper. Some of the specimens bear no similarity to plant and reptile bed material collected by the author since 2002, including the distinctive horsetail fragments assigned to *Equisetum mobergii* with their internodes and leaf teeth. Also not found in the plant bed are fossils comparable to the twig specimen assigned to *Brachyphyllum* sp. – despite abundant shoots and twigs from the bed, none have the leaf pattern of this morphogenus – and to the incomplete cone fragments assigned to the species *Araucarites sizerae*. This left only the seeds.

It took one brief look at the seed illustrations in the paper to realise their significance. The rewards of fossil collecting include an intense feeling of satisfaction on identifying unusual finds. Remarkably, of the seven seed species illustrated, five could be identified from the plant and reptile bed material collected by the author between 2002

---

192 Thomas and Batten (2001c), 132.

and 2006, and these finds were subsequently reported in 2007 in the *Wiltshire Archaeological and Natural History Magazine*.[193] Five seed species, with specimens preserved through a similar process of silicification, thus occurred in two different types of deposit found within different geological formations in south-central England, and to the author's knowledge at that time were unrecorded from any other localities in the world – but there did turn out to be one exception.

Two of the species from the plant and reptile bed, both with a length of about 3mm and with two contrasting surfaces meeting at the micropyle and hilum, match the described Portesham Quarry species *Carpolithes gibbus* and *Carpolithes acinus*. The micropyle is the opening in the integument or outer covering through which pollen enters for fertilisation, normally found at the tip of the pointed end of the seed, and the hilum is the stalk scar. Seeds with the micropyle at the opposite end to the base or hilum – a description that applies to all species mentioned in this chapter – are described as orthotropous. New species of fossils are described on the basis of a type specimen or holotype best exhibiting the key diagnostic features and kept in a recognised museum or other scientific collection, and the Portesham Quarry seed holotypes had been given to the Natural History Museum in London.

Neither of these seed species would take pride of place in a fossil collection on the basis of appearance. Both have a flat to slightly concave surface and a convex surface. In the case of the rarer of the two, *C. gibbus*, the name gibbus comes from the Latin for hump back and refers to the keel on the convex surface. In the case of the more abundant *C. acinus*, in which the flat surface is smooth and the convex surface striated, the name *acinus* comes from the Latin word for pip.[194]

Two further species of seed common to the plant and reptile bed and to Portesham Quarry's Charophyte Chert, *Carpolithes*

---

193 Needham (2007), 1–20.
194 Brown and Bugg (1975), 433.

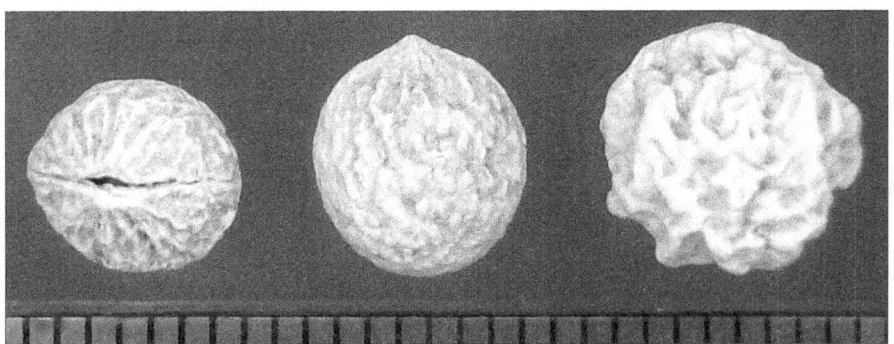

Silicified specimens of the seed Carpolithes rubeola from the plant and reptile bed, a species first described as a conifer from Portesham Quarry in Dorset. Note the specimen with micropyle uppermost and plane of separation clearly visible (left), and the specimen with an exceptionally rugged surface (right). Specimens: Author's collection. Photo: Author.

rubeola and *Carpolithes glans*, have two mutually perpendicular planes of symmetry passing through the micropylar and hilar ends, one of these lying along a plane of weakness such as that separating the two halves of a walnut or peach stone. A degree of asymmetry in some specimens could be due to irregular growth, but can usually be attributed either to irregular surface detail or to the effects of pre-fossilisation compression. These are two of the most abundant and most distinctive seeds from the plant and reptile bed.

C. rubeola was described on the basis of three specimens from Portesham Quarry, the holotype and two sectioned specimens. An 0.7mm thick stony layer of integument 'composed of three layers of stone cells'[195] was identified from the sectioned specimens, and the basic species diagnosis ran as follows: 'Seed orthotropous, stone broadly oval, rounded in section, apex abtusely pointed, base rounded: length 6 mm, width (major axis) 6 mm, depth (minor axis) 5 mm. Surface of stone shows both ridges and lumps.'[196] The apex here is the micropylar

---

195 Brown and Bugg (1975), 431.
196 Brown and Bugg (1975), 431.

end and the base the hilar end. Plant and reptile bed specimens have a size range that includes slightly larger specimens, and the quality and quantity of the specimens led to the plane of weakness being identified for the first time, and for a clear differentiation between structural ridges along this plane and apparently superficial ridges to be made.

Most remarkable of all the seed species collected from Portesham Quarry and also present in the plant and reptile bed must be *Carpolithes glans*, described by Brown and Bugg in 1975 on the basis of two specimens from Portesham Quarry, one of which was sectioned to reveal four layers of cells in the integument. The species name was taken from the Latin *glans* for acorn, because of a similarity in appearance to an acorn in a cupule. These two seeds were later compared with twelve specimens attributed to the species from the plant and reptile bed.[197] The latter specimens tend to be larger than those from Portesham Quarry, with overall sizes ranging from 4 to 7mm in length and 4 to 8mm in width and depth. Basic characteristics of the seed are a micropylar point at the apex, a grooved surface leading down to a lumpy basal ridge, and a flattened hilar or basal area within the ridge. A plane of weakness passes through the micropylar point, and along this plane there are two projections where it passes through the basal ridge. Some specimens retain a basal cover in varying states of attachment to these projections and the flattened basal area.

As with *C. rubeola*, the number and quality of the plant and reptile bed specimens allowed identification of the plane of weakness. More significantly, the projections on the basal ridge had not been noted in the Portesham Quarry diagnosis, and direct observation of the holotype reveals one of these to have been missing. The other cannot be discerned in the original illustration that accompanied the 1975 description, and could easily have been mistaken for a mineral

---

197 Needham (2007), 5.

growth. It was only the discovery of the projections and basal cover in *C. glans* that revealed the highly unusual nature of this seed species.

The collection of several specimens of *C. glans* during slow processing of the sediment provided information on the process of separation of the basal cover from the main seed body. A variety of reasons can be suggested for this process: examples include a response to fire leaving the seed ready for germination in a fertile ash bed with the arrival of the next rain, or a response to rain to allow water imbibition and germination. There may well have been other factors at play, and the ends of the projections on some specimens appear as hollow tubes that raise the question as to whether they had some other function. The possibility that the basal covers maintained the seeds in a state of physical dormancy has been considered,[198] although this would depend on whether they had been fertilised before dispersal – many fossil 'seeds' are in fact unfertilised ovules, with fertilisation occurring after dispersal. Whatever the truth may be, these seeds are one of the most fascinating finds from the plant and reptile bed and well worthy of further study.

Commonest of the seeds named by Brown and Bugg is *Carpolithes westi*, illustrated in Chapter One and named after Dr Ian West, who originally reported the material from Portesham Quarry and assisted in preparing the paper on it. 'A number of specimens' were reported from Portesham Quarry, and several dozen attributable or comparable to the species have been collected from the plant and reptile bed since 2002. Lengths of these wedge shaped fossils range from 4mm to 9mm and depths and widths from 2mm to 6mm. The holotype has four longitudinal ridges on one side, running from a narrow apex to a rounded or flattened base. The opposite side has 'a thickened median longitudinal region'.[199] Sectioning of specimens

---

198 Needham (2007), 5–6.
199 Brown and Bugg (1975), 433.

Specimens of the remarkable seed species Carpolithes glans from the plant and reptile bed. The basal cover can be seen in the process of separation in the specimens to the right, a feature not recorded in the original Portesham Quarry fossils. The attachment points are visible in the specimens top left and centre. Specimens: Author's collection. Photo: Author.

revealed similar internal structures interpreted as a nucellus, which is a type of tissue in the ovule, enclosing a membrane around a cavity with a tiny shrivelled body within.

It has already been recounted in Chapter One that this 'seed' was recognised as a possible worn araucarian cone scale, and that there were similarities with previously described or illustrated material. Suffice it to say at this stage that it completes the list of five seeds identifiable from both the Charophyte Chert of Portesham Quarry and the plant and reptile bed that had been collected by 2006, but that it will be met with again. Also worthy of a brief mention at this point is a very much larger specimen of a possible araucarian cone scale from the plant and reptile bed, 70mm long and with a maximum width of 24mm

Three other unnamed species of seed from the plant and reptile bed were briefly described in 2007 as Form A, Form B and Form C,[200] and although they are not recorded from Portesham Quarry this could well be due to the relatively small sample that was processed from that site. A number of other worn or indeterminate seeds were also collected, and one species from Portesham Quarry described on the basis of a single damaged specimen, *Carpolithes cocos*, has not been recorded by the author from the plant and reptile bed.

This leaves one Portesham Quarry seed species unaccounted for, *Carpolithes rhabdotus*. With over 160 fossil seeds already collected from the plant and reptile bed, the discovery of a single specimen of this species in 2008 came as something of a surprise. The species had originally been described and named on the basis of two specimens from Portesham Quarry, one of which was broken before study and subsequently sectioned.[201] Excluding any minor distortions, the seeds are radially symmetrical around an axis running from the micropyle to the hilum, and are 8 to 9mm in length and 5 to 7mm in diameter. The plant and reptile bed specimen has sixteen ridges running most of the way between the micropylar and hilar regions. A small conical projection on the basal or hilar end and 'a ring-shaped depression round a central raised area'[202] on the micropylar end were described on the holotype, and these features are readily discernible on the plant and reptile bed specimen.

A similar seed to *Carpolithes rhabdotus* had been described from older French Upper Jurassic deposits of Oxfordian age by G. de Saporta in 1875, but could be excluded as belonging to the same species as it was 'about three times larger than these [Portesham Quarry] seeds and it differs also in not being flattened around the hilum'.[203] Saporta

200 Needham (2007), 6.
201 Brown and Bugg (1975), 432.
202 Brown and Bugg (1975), 432.
203 Brown and Bugg (1975), 432.

had named the French seed *Cycadeospermum schlubergeri*, indicating that he regarded it as a cycadophyte seed, whereas the Portesham Quarry seeds were all placed in the Order Coniferales, but some of these attributions have to be seen as somewhat tenuous.

It is regrettably difficult to interpret the features of petrified seeds from plants of unknown size, habitat and classification that grew around 146 million years ago. With a living plant, seed development and germination can be witnessed within the environment to which it is adapted, but fossils are a different matter. *Carpolithes rhabdotus* received the species name *rhabdotus* from the Greek word for fluted, owing to the rounded ridges and furrows on the seed's surface. Such ridges could assist the seed in anchoring itself to the soil, but could equally allow it to expand following the take up of water after rain. This is the only sub-spherical seed species from Portesham Quarry or the plant and reptile bed, and the raised feature at the micropylar end could be an example of a 'plug-like' type of seed appendage 'that separates during germination by circumscissile dehiscence',[204] or in other words via the opening of a circular fissure. 'Operculum', 'strophiole' and 'chalazal cap' are the names for various small structures in the water-impermeable layers of today's seed coats that move or become dislodged to allow the entry of water,[205] and in the possibly semi-arid,

The single specimen of the seed species Carpolithes rhabdotus collected by the author from the plant and reptile bed, showing the fluted surface and small cone on the base. The species was originally described as a conifer in 1975, although a similar but larger seed from France had been interpreted as a cycadophyte seed. Specimen: Author's collection. Photo: Author.

204   Fahn (1990), 513.
205   Baskin and Baskin (1998), 39, Fig. 3.2.

seasonally unstable climate postulated for the Purbeck Fossil Forest the occurrence of similar seeds which must germinate rapidly upon the arrival of seasonal rainfall is a credible scenario. The possibility that forest fire could have played a part in stimulating germination can also be considered.

The discovery of *Carpolithes rhabdotus* raised to six the number of seed species common to the Charophyte Chert and the plant and reptile bed, and further significant finds were to follow between 2008 and 2010. The first of these finds took the form of four specimens that offered some potentially new and significant information on *Carpolithes rubeola*. The most complete and finely preserved was a large egg-shaped 'seed' some 14mm long from the micropyle to the hilum, 11mm across at the broadest point, generally smooth-surfaced but featuring a pair of opposite ribs running away from the micropyle and developing into truncated ends near the hilum. Next was an apparent seed shell 11mm long and 10mm across at the broadest point, about 1mm thick and featuring the same basic surface features as the previously described specimen, albeit in a less well-preserved state. The shell was hollow and split open along one side. So far there was nothing to connect these specimens with *C. rubeola*, and they indicated a nut-like species in which the kernel in one case had either rotted away or been eaten.

A third specimen confused the matter. It was a split half of a 'seed' 13mm long and 11mm across of which the missing half had broken away and was lost during collecting. The micropyle and one faint ridge were visible on the surface, and the split through the centre revealed a hollow the approximate shape of a *C. rubeola* seed that had partly infilled with calcite crystals. The outer layers of the seed were visible in cross section, but not crinkled as in *C. rubeola* seeds found in isolation. At this stage it seemed likely that *C. rubeola* seeds developed within an outer shell, within which an inner shell shrank and crinkled before dispersal.

A fourth specimen, 12mm long and 11mm across, consisted of an apparently degraded outer layer of which one part was missing, revealing just over half of the shell or integument of a *C. rubeola*-shaped seed. The inner parts of the seed were absent. This remarkable specimen, apparently belonging to a species attributed by Brown and Bugg to the order Coniferales, is superficially at least comparable to seeds of the order Ginkgoales. In these seeds the seed coat or hard outer part of the integument is called the testa or sclerotesta, and the outer layer is called the sarcotesta – this layer can be fleshy or leathery, but for fossilisation to take place would in all but the most exceptional of conditions need to be the latter. Although seeds of the extant maidenhair tree *Ginkgo biloba* are larger than those of *C. rubeola* and smooth surfaced, examples of these seeds both removed from and within their sarcotestas invite an immediate comparison, as does a Palaeocene *Ginkgo* seed from North Dakota with the sarcotesta partially preserved.[206] The Palaeocene is the earliest epoch of the Tertiary, and ginkgophytes apparently peaked in terms of diversity and distribution during the late Mesozoic and early Tertiary when 'they achieved a circumpolar distribution in the Northern Hemisphere and extended into several regions of the

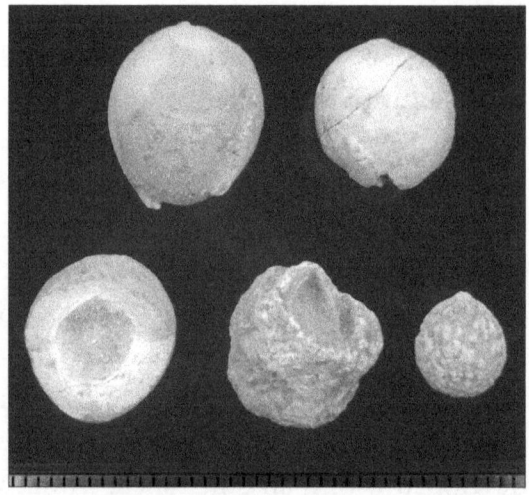

Fruits from the plant and reptile bed enclosing seeds of the species Carpolithes rubeola (top, bottom left and centre) and an isolated seed (bottom right). Features of this fruit are comparable to those of extant and extinct members of the ginkgo family, and ingestion by herbivores could have been an effective means of seed dispersal. Specimens: Author's collection. Photo: Author.

206 Royer, Hickey and Wing (2003), 85, Fig. 1B.

Southern Hemisphere'.[207]

Already well represented by seventeen morphotaxa in the Middle Jurassic of Yorkshire, the occurrence of ginkgophytes in a British Tithonian flora would thus be no surprise. Recent evidence suggests that Mesozoic ginkgophytes lived in a far broader range of environments than the modern maidenhair trees, with charcoal ginkgophyte branch remains from the Upper Cretaceous of the Czech Republic interpreted as forming, along with 'cheirolepidiaceous conifers', a 'fire-prone vegetation in halophytic salt marsh environments under a seasonal, sub-tropical climate'.[208] A halophyte is a plant adapted to growing in salty conditions, and comparisons between some aspects at least of this Czech environment and that of the earlier Purbeck environment are easily discernible. There are other groups of Mesozoic seed-bearing plants to which *C. rubeola* could belong, and the correct answer will no doubt be derived from detailed scientific investigations.

Whatever its taxonomic affinities, a possible picture emerges of *C. rubeola* as a hard stone enclosed in a tough sarcotesta. The degraded specimen could contain the remains of a seed that had germinated, and in the maidenhair tree such seeds with a degraded sarcotesta germinate more successfully than those with an undegraded sarcotesta.[209] Most likely to germinate could well have been seeds in which the sarcotesta had completely gone, and the seeds of the maidenhair tree can be spread by badgers which eat them, digest the sarcotesta and pass the seed in their droppings, whence they germinate. In the Upper Jurassic similar seeds could have been spread by herbivorous dinosaurs – indeed, according to one suggestion, 'the odoriferous sarcotesta of *Ginkgo* would have been well adapted to attracting

---

207 Rothwell and Holt (1997), 223.
208 Falcon-Lang (2004), 349.
209 Rothwell and Holt (1997), 226.

herbivorous reptiles'.[210] This would suggest selective feeding, although it has been claimed that the larger-bodied herbivores would have eaten indiscriminately, because they required so much bulk that they 'cannot subsist using selective feeding processes'.[211] Whatever the case, it is easy to imagine herbivores of which some species are believed to have fed on tough, woody vegetation swallowing the whole fruit and digesting the outer sarcotesta. The hard, stone-like sclerotesta could then have been adapted to survive briefly in a gizzard part-filled with gastroliths or grinding stones before being passed through to a new germination site.

Seed-eating Jurassic mammals would mostly have been too small to eat whole a sarcotesta some 13mm in length, but a possible fate for the seed with the apparently degraded sarcotesta – assuming that it had not germinated or was not simply an example of incomplete silicification or mineralisation– is that a small mammal had cut through what was probably a tough leathery coating and gone for the more nutritious centre. Suspicion could fall on a multituberculate with incisors that 'probably were used to winkle out endosperm from seeds that were first crushed by the bladed pre-molars'.[212]

Four fossils collected between 2002 and 2006 from the plant and reptile bed were originally identified as seeds comparable to an American seed genus *Jensensispermum*.[213] These specimens were donated to the NHM, but three further finds by the author have led to a re-evaluation of the material. They have two mutually perpendicular planes of symmetry passing through an axis running from a lipped end that was formerly interpreted, on the basis of the original description of *Jensensispermum*, as the micropylar end, and an opposite closed end interpreted on the same basis as the hilar end. One of the new

210 Rothwell and Holt (1997), 227.
211 Taggart and Cross (1997), 406.
212 Benton, Hooker and Cook (2005), 59.
213 Needham (2007), 6–7.

specimens was typical of earlier donated finds, with a length of 7mm, a width of 5mm and two fine grooves running back some two-thirds of the way along the specimen from above each lip. Another specimen of similar size had part of one side missing, exposing one of what was clearly a pair of chambers sitting either side of a dividing partition running between the two ends. The third specimen, once again with a length of about 7mm but with a width of 6.5mm, appeared to have opened out laterally, revealing twin channels running back from the lips on each side of the micropylar end and separating before tunnelling through to the interiors of the chambers. Without cutting open a specimen or scanning it, it would appear that a channel from each pair ended in adjacent openings into the seed chamber, visible in the specimen with part of one side missing. The pairs of fine grooves running back from above each lip had now developed into distinct cracks in this third specimen, running most of the length of the fossil.

One interpretation of these specimens is that they are capsules enclosing two seeds, with a pair of channels for airborne or water droplet pollination entering each chamber. As the capsule expands the channels are exposed and pairs of cracks over each chamber prepare to open before seed dispersal. This leaves one very obvious question – are there any seeds from the plant and reptile bed that fit into the chambers? There is in fact one very suitable candidate, in the form of *Carpolithes acinus*. With the flat side against the partition, these seeds fit in well. Furthermore, a close examination of examples in the

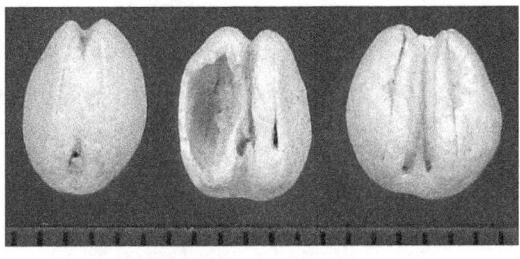

Probable seed capsules from the plant and reptile bed, each of which would have enclosed two seeds, possibly of the species Carpolithes acinus. These specimens have been compared with the American seed genus Jensensispermum, a possible cycadophyte, but they could be from a pteridospermophyte or Mesozoic seed fern. Specimens: Author's collection. Photo: Author.

author's collection reveals three specimens in which the micropyle is situated away from the flat side of the seed and some way up the striated convex side, which would place it adjacent to the channel openings in the chamber.

The capsules are not characteristic of conifer seed-bearing structures, and are more readily comparable with cupules that almost completely enclose seeds in some of the Mesozoic seed ferns. Some caytonias such as *Caytonia* from the Middle Jurassic of Yorkshire, originally interpreted as early angiosperms, have cupules enclosing several seeds, but *Ktalenia* from the Lower Cretaceous of Argentina 'has only two seeds per cupule'.[214] Evidence suggests that some species of the associated foliage morphogenus *Ruflorinia* lived in xeric or arid environments. Cupules of the Triassic corystosperm *Umkomasia resinosa* contain 'either one or two ovules'.[215]

Examples of fossil wood from the plant and reptile bed tend on average to be far smaller than specimens from the Purbeck Fossil Forest, which are mostly the remains of large branches, trunks and stumps. A wood specimen from the plant and reptile bed with a length of about 55cm, collected in 2007, was one of a few notable exceptions. Such specimens sometimes show evidence of wear typical of erosion by moving water and are confined to the base of the bed. One of these, 30cm long and from 5cm to 7.2cm in diameter, was transversely sectioned at the Natural History Museum before donation to their collections. This revealed moderate anatomical preservation, with conifer cell walls clearly visible along with irregularly spaced growth rings. The ring sequence invites comparison with a sequence from Purbeck Fossil Forest wood illustrated in *Jurassic Coast*.[216] There is a clear shared pattern of irregular spacing, which probably includes true annual rings and 'false' or 'intra-annual rings' seen as 'indicative

---

214 Taylor, Taylor and Krings (2009), 626.
215 Taylor, Taylor and Krings (2009), 634.
216 Brunsden (ed.) (2003), 50.

of the onset of water shortage during the growing season'.[217] It was comparisons between similar fossil growth ring patterns and those on today's trees that led Francis to interpret the early Purbeck climate as Mediterranean with 'conditions marginal for tree growth and highly irregular from year to year'.[218]

The Portesham Quarry flora included many conifer wood fragments and 'a good many less characterized specimens which are omitted and also obscure bodies of uncertain nature'.[219] 'Obscure bodies of uncertain nature' is a good description of some of the specimens from the plant and reptile bed, and it would be interesting to know if there is any further material common to both sites yet to be uncovered.

Viewed overall, the plant and reptile bed seed and wood assemblage provides solid evidence that an environment similar to that of the basal Purbeck conditions in the west of Dorset existed on nearby land in or adjacent to what is now south Wiltshire during Portland Stone Formation times. Two at least of the six seed species confirmed as being common to Portesham Quarry and the plant and reptile bed show evidence of specialised mechanisms that are likely to be environment-specific, even if that environment cannot be defined. If, as is very possible, fires occurred within the forests or scrubland in which they grew, they could have been adapted to this in different ways – some to survive as part of a seed bank in the soil, some protected by hard coats, some to be activated by fire-induced mechanisms and some to be brought in by the wind or animals from undamaged areas. Added to this are possible mechanisms of drought survival and of water absorption before optimally timed germination. Whatever the answers, the seeds of the plant and reptile bed should provide some truly remarkable insights into Upper Jurassic palaeocarpology.

217 Francis (1984), 296.
218 Francis (1984), 285.
219 Brown and Bugg (1975), 427.

A carapace fragment of a turtle of the extinct freshwater family Solemydidae from the plant and reptile bed. This family, readily identifiable from the tubercles on the surface, was previously only recorded from Cretaceous strata, and finds from Wiltshire's Tithonian deposits thus extend its known range into the Upper Jurassic. Specimen: NHM collections. Photo: Author.

Some of the reptile finds common to the plant and reptile bed and the Purbeck Limestone Group have already been mentioned, and a brief overview of taxa common to both assemblages follows. Where applicable, references to these finds can be found in the preceding text. Before going into the reptiles, it should be noted that one small bone collected by the author from the plant and reptile bed has been identified as possible frog.[220] This is a potentially notable addition to the English Tithonian-Berriasian frog record, previously confined to examples described in Chapter One – an unconfirmed 1876 Wiltshire report and a species from the Purbeck Limestone Group of Sunnydown Farm in Dorset. The bone also shares key features with the ribs of some choristoderes, a group of aquatic lizards with a fragmentary fossil record.

Turtles were originally unreported from the plant and reptile bed in published SSSI and GCR faunal assemblages, but two small fragments of turtle shell collected by the author in 2010 have been identified by the NHM as belonging to the Family Solemydidae, which occurs in the Purbeck Limestone Group of Dorset in the form of the genus *Helochelydra* (see Chapter One). Solemydids have

---

220 Barrett, P. (2010) Pers. comm.

'distinctive enamel-covered tubercles on the external shell surface', and are 'a poorly known family of fresh water turtles described from the Upper Cretaceous of North America and the Iberian Peninsula and the Lower Cretaceous of Great Britain and Texas'.[221] The Lower Cretaceous British specimens have been found in Purbeck and Wealden strata, and in the case of the Purbeck Limestone Group of Dorset are known from various horizons between the Mammal Bed near the top of the Lulworth Formation and the Crocodile Bed in the Durlston Formation.[222] Approximations based on frequency of specimens suggest that *Helochelydra* comprised only about 3 per cent of the turtles in the Purbeck Limestone Group, and along with other infrequent forms were either 'rare elements in the fauna or transported transients/erratics'.[223] The solemydid fragments from the plant and reptile bed are thus significant on two counts, firstly because at a global level they take the family record clearly back into the Upper Jurassic, and secondly because they are a group that links various horizons across the Upper Jurassic and Lower Cretaceous fresh water/terrestrial interlude in southern England. A further turtle find was made in 2010 in the gastropod micrite, an as yet unidentified non-solemydid specimen consisting of numerous carapace fragments within a block of limestone.

Crocodile teeth attributable to the genus *Goniopholis* and to the atoposaurid genus *Theriosuchus* and comparable to the genus *Bernissartia* have been reported from the plant and reptile bed and collected from it by the author, and occur in various Purbeck Limestone Group beds in Dorset. Pterosaur material collected by the author is indeterminate, but the pterosaur genus *Gnathosaurus* has been reported from the plant and reptile bed and is one of the relatively large filter-feeding ctenochasmatids described from the

---

221 Bennett (2008), 52.
222 Milner (2004), 1444.
223 Milner (2004), 1465.

Dorset Purbeck Limestone Group. No specific sauropod taxa have been identified as common to the plant and reptile bed/gastropod micrite and the Purbeck Limestone Group of Dorset, although evidence of them is common to both strata and teeth identified as or comparable to the probable brachiosaurid genus *Pelorosaurus* have been found by the author in the gastropod micrite and historically in the Purbeck Limestone Group of Buckinghamshire. Of the carnivorous theropods, the 2005 GCR brief reference to 'small and large' forms from the plant and reptile bed is reinforced by the author's discovery of small teeth from the plant and reptile bed and dromaeosauroid and allosauroid specimens from the gastropod micrite – these indicate a comparable fauna to the Purbeck Limestone Group of Dorset with its dromaeosaurid genus *Nuthetes* and an *Allosaurus*-like tooth, but the problems involved in making specific identifications from isolated teeth have already been highlighted. Two specimens collected by the author from the plant and reptile bed have been identified by the NHM as 'cf. *Nuthetes*'. Other common elements in the reptile fauna occur only at family or less determinate level – iguanodonts in the form of reported *Iguanodon* teeth from the plant and reptile bed and *Camptosaurus* from the Purbeck Limestone Group of Dorset, and possible nodosaur teeth from both strata.

A small fragmentary fossil with a surface area of about 6sq.mm collected by the author from the plant and reptile bed was identified as probable crocodile eggshell. The identification was made at the NHM by Paul Ensom, who had reported fragmentary dinosaur, crocodile and turtle eggshell from the Cherty Freshwater Member of the Purbeck Limestone Group in 1996, providing 'unequivocal evidence of the United Kingdom's first dinosaur eggs'.[224]

The only snippet of information about the plant and reptile bed mammals, i.e. that members of three orders, Multituberculata,

---

224 Ensom (1996), 79.

Triconodonta and 'Eupantotheria', had been positively identified,[225] indicates a broadly similar fauna to that of the Purbeck Limestone Group. Finds of multituberculates have already been noted from Swindon and Durlston Bay in the previous chapter, and triconodonts from Durlston Bay. 'Eupantotheria' includes the dryolestids, of which five species from Durlston Bay were referred to as cladotheres in the previous chapter. The classification hierarchy of these early mammals is increasingly complex, and similar species are often referred to under alternative group names by different sources. Suffice it to say for the purposes of this description that a number of probable seed-eating, insectivorous and carnivorous small mammals lived in Tithonian-Berriasian terrestrial habitats in Wiltshire and Dorset.

When viewed overall, and taking into account the many gaps, the floral and faunal assemblages of the plant and reptile bed

Teeth of the crocodile Goniopholis from the plant and reptile bed (left) and the Cherty Freshwater Member of the Purbeck Limestone Group (right). Specimens of this genus are abundant from both horizons and indicate common environmental conditions in Tithonian south Wiltshire and Berriasian Dorset. Specimens: Author's collection. Photo: Author.

225 Benton, Hooker and Cook (2005), 54-6.

and gastropod micrite suggest that they originated from terrestrial environments similar to those that prevailed during deposition of the Purbeck Limestone Group. As such they represent a marine regression during Portland times with a Purbeck-type facies. Facies refers to the defining characteristics of a rock formation, in this case with specific emphasis on fossil content rather than lithology – the sandy silt is probably a residue from erosion of sandy Tisbury Member limestones, and thus distinct from Dorset Purbeck Limestone Group deposits.

One factor that needs to be considered in looking at the Portland-Purbeck deposits of Dorset and Wiltshire is the imperfectly understood chronostratigraphy. The Portland Stone Formation and the Purbeck Limestone Group are lithostratigraphic units, or units of geological strata defined by their rock characteristics. Chronostratigraphic units are units of geological strata defined by the time period in which they were laid down. Although lithostratigraphic units *can* co-equate with chronostratigraphic units where widespread conditions of deposition change simultaneously, possibly as the result of a sudden major tectonic movement or change in sea level, there are other conditions such as a slow marine regression or localised tectonic movements where the change from one type of sedimentation to another occurs at different times across a large area. In the case of the Portland-Purbeck transition this is a possible scenario.

At Portesham Quarry the change from Portland to Purbeck facies is reflected more in fossil content than in lithology. The fossils suggest a fluctuating salinity at the very base of the Purbeck Limestone Group as stable marine conditions came to an end. In the Vale of Wardour the Portland-Purbeck transition is more straightforward at sites such as Upper Chicksgrove Quarry, where 'Unmixed Purbeck facies directly overlies the Wockley Member micrites ... with a gradational contact'.[226] At Chilmark Ravine, however, 'The junction

---

226 Wimbledon (1976), 6.

between the oolites and the basal Purbeck Beds is not a clear cut one ... Portland and Purbeck facies are mixed.'[227] At the northernmost Wiltshire outcrops at Swindon 'the Purbeck Beds ... interdigitate with marine beds of Portland facies', and 'we must conclude that the basal Purbeck Beds in some places were being laid down at the same time as the upper Portland Beds in others'.[228] A possible interpretation is that there was an earlier marine regression in north Wiltshire than in Dorset, leaving open the possibility that the plant and reptile bed was deposited at not too different a time. The thickness of intervening deposits has to be taken into account here, but rates of deposition vary according to local conditions and regressions are accompanied by levels of erosion that also vary from place to place. These problems have been the subject of considerable study and their resolution will most likely occur through correlation of invertebrate marine, brackish and freshwater fossil assemblages. Unfortunately, after several years in preparation, publication of a long-awaited GCR volume on the Jurassic-Cretaceous transition in southern England was cancelled in August 2011. Meanwhile, a recent examination of a number of invertebrate fossils, mostly gastropods, from the plant and reptile bed by Dr Roy Clements[229] has identified species derived from terrestrial, freshwater/brackish and marine conditions. The assemblage has been compared with material from Dorset basal Purbeck sites, including Portesham Quarry, and Purbeck strata at Swindon, and it once again emphasises the close palaeontological relationship between the plant and reptile bed and the basal beds of the Purbeck Limestone Group.

The geographical, or rather palaeogeographical, context of this deposition in Dorset and Wiltshire has been reconstructed over recent decades in a way that was not possible for nineteenth-century geologists. Boreholes drilled across southern England for oil

---

227 Wimbledon (1976), 6.
228 Barker, Brown, Bugg and Costin (1975), 422.
229 Clements, R. (2009) Pers. comm.

The author's daughter Nicole Needham observes the uncovering of a 55cm long piece of silicified wood. She is seated on the gastropod micrite, her feet are resting on silt of the plant and reptile bed, and the bottom of the wood is resting on the top of the Tisbury Member marine sandy limestones. Specimen: Author's collection. Photo: Author.

exploration purposes have provided a mass of data on the basins in which sediments were forming, and on the basis of some eighty boreholes detailed maps of the thicknesses of geological deposits during the entire Jurassic period have been made.[230] Marine deposition over most of south-central and south-east England persisted throughout the period, centred by Portland Stone times on two connected basins named the Weald Basin and the Wessex Basin. These in turn connected southwards via a Channel Basin to what is now northern France and the Paris Basin. As the Portland Stone was being deposited in south Wiltshire, wood and plant debris were deposited as 'features of the shallow marine rocks ... showing the proximity of land to the south, over a north Dorset swell, and/or to the north of the Mere fault'.[231] The Mere fault, named after the large village of Mere some 7 miles west-north-west of Tisbury, was an active fracture in the earth's surface along the north-west margin of the Wessex Basin, to the south and south-east of which the rocks were moving downwards. These tectonic movements could well

---

230 Sellwood, Scott and Lunn (1986).
231 Cope, Rawson and Wimbledon (1992), 124.

have been associated with the early stages in the opening up of the Atlantic Ocean, but with uncertainty over the direction in which the nearest land lay any detailed reconstruction of Tithonian topography in south Wiltshire is conjectural. For much of the Jurassic period the entire area now covered by the British Isles formed part of an island archipelago, with intermittently connected landmasses centred on the south-west peninsula, Wales, Scotland, the Hebrides, eastern England and the Low Countries, and Brittany. The Purbeck Limestone Group was deposited at a time when southern England was situated 'at a palaeolatitude of about 36°north', the current latitude having been reached by northward continental drift. During Tithonian times it is believed that many of these landmasses became connected to each other as sea levels fell, producing continuous land connections around the increasingly enclosed Wessex-Weald-Channel Basins from Brittany to south-west England, Wales, eastern England and the Low Countries.[232]

Before discovery of the plant and reptile bed a broad picture of Portland marine conditions and of an ensuing marine regression leading to Purbeck conditions in Dorset and Wiltshire had been constructed. Discovery of the plant and reptile bed revealed evidence of an earlier marine regression, of unknown duration but probably relatively brief in geological terms, that was accompanied by the emergence of land extending either to the south or to the north of what is now south Wiltshire. Erosion and lagoon formation occurred before the re-establishment of marine conditions. Evidence from the flora, the invertebrate fauna and the vertebrate fauna of the plant and reptile bed repeatedly indicates that conditions from which this material originated were very similar to those of the ensuing Purbeck times. Vertebrates from the plant and reptile bed, coming as they do from below the base of the Purbeck Limestone Group, point to

---

[232] Cope, Rawson and Wimbledon (1992), 129, Maps J11a, J11b, J11c and J11d.

a continuity of conditions upwards to the higher horizons of the Lulworth Formation, and support the proposal that the vertebrate assemblages of the Marly and Cherty Freshwater members represent similar creatures to those that lived during the time in which the Purbeck Fossil Forest was formed.

The regional picture of the Tithonian world of south-central England becomes a little less blurred, but the finds from the plant and reptile bed extend the picture far beyond the confines of Europe. When Brown and Bugg described the land plants from Portesham Quarry they had few other descriptions of Upper Jurassic seeds to refer to, but they gave one reference. This was to a paper written by Marjorie Chandler in 1966 on some plant remains from the American state of Utah, and the story of the plant and reptile bed was about to enter a new dimension.

4
FROM WILTSHIRE TO THE AMERICAS

ABOUT A DECADE before publication of Brown and Bugg's description of the land plants from Portesham Quarry, an American couple named Homer and Joanne Behunin from the town of Redmond in Utah discovered an assemblage of fossil plant remains at a locality referred to as the Fremont site, situated some 5 miles east of Willow Springs, also in Utah. The specimens were three-dimensionally preserved 'crystalline casts of external moulds of plant remains',[233] and the geological strata from which they were collected lay within the Brushy Basin Member of the Morrison Formation, described by James A. Jensen, Curator at the Geology Department of Brigham Young University, Provo, Utah, as 'in the upper part of the Jurassic, if not at its very top'.[234] Jensen worked with Homer Behunin at the site and was put into contact with Marjorie Chandler of Isle of Sheppey fame.

Since co-writing *The London Clay Flora* with Reid in 1933, Chandler had further consolidated her international reputation in

---

233 Chandler (1966), 143.
234 Jenson in Chandler (1966), 141.

Silicified plant fossils from the Henry Mountains of Utah. Similar specimens collected from the Fremont site in Utah were initially interpreted as cycadophyte seeds by Chandler under the name Behuninia joannei. Many such fossils are mineral casts, but some are petrifactions preserving internal structure. Specimens: Author's collection, collected Richard Dayvault. Photo: Author.

palaeobotany and had continued to focus her research on Tertiary angiosperms. She was now invited to examine a far older fossil assemblage, and in her own words, 'The sole justification, therefore, for attempting a study of these older fossils is that in their superficial appearance some of them bear a misleading resemblance to Angiosperms, while their preservation as solid entities, can be interpreted in the light of many years of research on similarly preserved Tertiary fruiting organs.'[235] She was not to be without some support from a Mesozoic specialist of unrivalled reputation, however, as Professor Tom M. Harris examined key specimens and offered opinions. They came to a similar conclusion regarding the Utah plant fossils, as expressed by Chandler: 'The best characterized and most abundant forms from Utah are not only entirely new to me but to Professor Harris also.'[236]

Unfortunately, despite the unusual three-dimensional preservation, the internal structure of the plant remains was not

---

235 Chandler (1966), 141.
236 Chandler (1966), 141.

preserved. As internal structure is one of the most important features used in the identification of such fossils, Chandler was working with one hand tied behind her back as she embarked on the daunting task of interpreting them. In the resulting paper, *Fruiting Organs from the Morrison Formation of Utah, U.S.A.* published in 1966, she described five new plant species attributed to three new genera and one existing genus. Type specimens were placed in the Brigham Young University collection and further specimens donated to the Natural History Museum in London.

Meanwhile, however, someone else had been collecting similar plant fossils from the Morrison Formation. In 1964 Charles Bass wrote a brief description of specimens collected from Mt Ellen in the Henry Mountains of south-eastern Utah, making no formal identifications but observing that the fossils were 'suggestive of both angiosperm and gymnosperm fructifications'. The published manuscript was entitled *Significant new fossil plant locality in Utah*,[237] and 'significant' would have been an understatement had the site yielded verifiable Jurassic angiosperm fructifications. Chandler added a postscript to her paper, noting that the descriptions by Bass had come to her attention.

The fossil plant remains described by Chandler were all attributed to the cycadophytes and other gymnosperms or deemed indeterminate. One new genus and species, *Behuninia joannei*, named after the Behunins, was placed in Family Cycadales (now usually Order Cycadales) of the Cycadophyta and interpreted as the seeds and the rachis or seed stalk of a 'true cycad'. Isolated seeds of another new genus and species, *Jensensispermum redmondi*, named after the Brigham Young University curator and the home town of the Behunins, were not deemed determinate to family level but provisionally referred to an unnamed and previously unknown family of the Cycadophyta.[238]

237   Bass (1964).
238   Chandler (1966), 156.

Fossils attributable to the Coniferales included cones interpreted as *Sequoia* sp., the genus which includes today's sequoias, and cones of a new genus and species, *Hillistrobus axelrodi*. The latter cones were provisionally referable to the same family as the sequoias, the 'Taxodineae' (Sub-family Taxodineae is now usually designated Family Taxodiaceae). All that remained were some seeds attributed to the morphogenus *Carpolithus*, not to be confused with the morphogenus *Carpolithes*, and placed under the general heading '*Incertae Cedis*' or indeterminate. There were two new species, *Carpolithus provoensis* and *Carpolithus radiatus*, as well as some mostly broken or compressed seeds referred to as *Carpolithus* sp.

Examples from Mt Ellen of the fossil types now described by Chandler as '*Behuninia, Jensensispermum* and *Carpolithus radiatus*',[239] had been illustrated by Bass in 1964, and there seems to have been common ground between Bass and Chandler in interpreting the fossils as seeds or fructifications. When illustrations of Chandler's Utah specimens were compared with plant and reptile bed material from Wiltshire around forty years later, specimens of two species, *Behuninia joannei* and *Carpolithus radiatus*, were apparently common to both sites. There were also remarkable shared features between one specimen in particular of the seed species described by Chandler as *Jensensispermum redmondi* and a few seeds from the plant and reptile bed, later interpreted as possible seed capsules of *Carpolithes acinus* (see preceding chapter). A search through the relevant literature, meanwhile, would reveal that further research had led to considerable revisions of the American flora.

In 1990 Professor William D. Tidwell of Brigham Young University wrote an overview of the megafossil flora of the Morrison Formation, in which he pointed out that the plant remains were mostly casts and petrifactions. Of a formation with a greater abundance of

---

239 Chandler (1966), 171.

From Wiltshire to the Americas

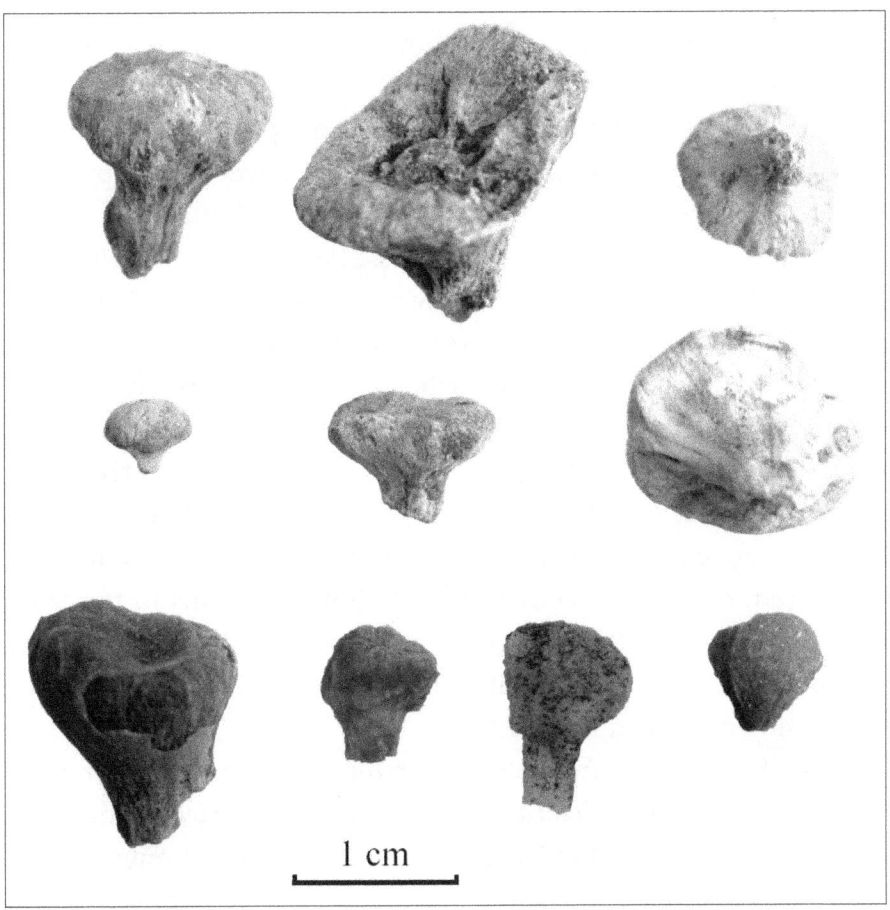

Specimens such as these from the plant and reptile bed (top and middle rows) and the Henry Mountains of Utah (bottom row) are now attributed to the probable short shoot species Steinerocaulis radiatus, but the first specimens from Utah described in 1966 were interpreted as seeds under the name Carpolithus radiatus.
Specimens: Author's collection, Utah specimens collected Richard Dayvault.
Photo: Steve Clifford.

petrified wood than any other in the USA, Tidwell could state that 'little research or attention has been focused on the fossil vegetational aspects'.[240] Referring to what research had been carried out, he reported

240 Tidwell (1990), 1.

a number of petrified ferns including tree ferns from Wyoming and Utah, cycadeoid (bennettite) trunks of the genus *Cycadeoidea* from Wyoming, South Dakota, Colorado and Utah, possible ginkgophyte seeds from the Fremont site in Utah, a range of compression fossils including leaves of several groups from Montana, and most abundantly petrified conifer remains. These latter, largely from Utah, included wood of the genera *Araucarioxylon*, *Xenoxylon* and others, twigs of *Brachyphyllum* with adpressed scale-leaves, cones of the Taxodiaceae as described by Chandler, araucarian-like cones, and short shoots. Of particular note was Tidwell's remark on Chandler's fossils from the Fremont site: 'Many of these structures, such as *Behuninia* and *Carpolithus*, have been subsequently found structurally preserved and thin-sections have shown them to be weathered, coniferous short shoots.'[241] Some revisions to the work of Chandler were clearly called for, and Tidwell and co-author David A. Medlyn of Utah State University obliged with an updated description published in 1992.

In *Short shoots from the Upper Jurassic Morrison Formation, Utah, Wyoming, and Colorado, USA* Tidwell and Medlyn drew on new material with anatomical structure preserved and reinterpreted Chandler's material. The 'seed' genus *Behuninia* survived in an amended form, now a morphogenus of 'short shoots probably of extinct conifers', and a further new probable short shoot genus *Steinerocaulis* was proposed.[242] Short shoots occur in plants with long shoot-short shoot organisation, and extant examples among the Gymnospermophyta are to be found in some conifers of the pine family and in the maidenhair tree *Ginkgo biloba*. In this type of organisation the main growth is provided by the long shoots, whereas the short shoots are generally stubby lateral structures usually crowded with

---

241 Tidwell (1990), 5.
242 Tidwell and Medlyn (1992), 219.

leaves growing from axillary buds. Short shoots are also sometimes referred to as spur shoots, although 'a spur shoot tends to have some internodal elongation, hence its distinction from short shoots, which lack internodal elongation'.[243] In many references to these Upper Jurassic fossils the terms have been used interchangeably.

The internal structure of the short shoots took the form of a small central pith and of radiating secondary xylem. The outer tissues of the cortex were generally absent. All the species described possessed these common internal features, and they were thus differentiated one from the other on morphology or external shape. Chandler's *Behuninia joannei* was the only species to retain its binomen or full scientific name intact, with most specimens now diagnosed as short shoots with pointed apices oppositely to suboppositely attached to axes – in other words, pointed shoots growing in pairs on stalks. Some specimens attributed to this species by Chandler were reallocated to a proposed new species of more rounded shoots under the name *Behuninia scottii*, named after a couple from Wyoming who assisted in the study. A few others were placed with new material and with Chandler's *Carpolithus provoensis* 'seeds' in the short shoot species *Behuninia provoensis*. This species incorporated detached shoots, 'lanceolate- to subglobular-shaped, often tapering towards each end'.[244] A fourth species was named *Behuninia bassii*, after Charles Bass who first reported the Mt Ellen site ahead of Chandler's paper. The diagnosis for this species was based on whorled to subwhorled attachment of the shoots, which otherwise shared many features in common with *B. provoensis*. Finally, Chandler's distinctive peltate or shield-shaped specimens of the 'seed' species *Carpolithus radiatus* were placed with newly introduced material into the second short shoot genus and renamed *Steinerocaulis radiatus*, after Richard Steiner

---

243   www.botgard.ucla.edu/botanytextbooks/generalbotany/types (visited 2007).
244   Tidwell and Medlyn (1992), 228.

Long shoot – short shoot organisation in living (left) and dead (centre) larch branches of the genus Larix, Family Pinaceae, and fossil short shoots (right) from the plant and reptile bed. The upper specimen is comparable to specimens of Behuninia joannei from Utah, which have been interpreted as belonging to needle-bearing conifers. Specimens: Author's collection. Photo: Author.

of Wyoming for his help with the study and the radiating structure. Most specimens were found detached, but where attached to an axis the manner was 'spiral to subspiral ... occasionally opposite'.[245]

Before the plant specimens collected by the author between 2002 and 2006 from the plant and reptile bed had been compared with the illustrations and descriptions by Tidwell and Medlyn, many had already been identified from a later source. Nonetheless, retrospectively a few specimens from the plant and reptile bed can be identified as comparable to *Behuninia joannei* and several more as all but certainly belonging to *Behuninia provoensis* and *Steinerocaulis radiatus* as described by Tidwell and Medlyn. A remarkable picture was emerging. A previously undescribed Wiltshire flora incorporated

245 Tidwell and Medlyn (1992), 232.

substantial common elements from both the Purbeck Fossil Forest horizon of Dorset in terms of seeds and the Brushy Basin Member of the Morrison Formation in terms of shoots.

In 1998 an illustrated interpretation of *Behuninia* shoots was made by Tidwell in a book on fossil plants from the west of North America. The illustration is of a main axis or branch with truncated long shoots bearing short shoots densely covered in pine-like needles. Evidence for the needles was based on tiny indentations on some specimens, which were interpreted as conifer needle scars.[246]

In 2003 an article entitled 'Short shoots from the Late Jurassic Morrison Formation of Southeastern Utah' was published in the magazine *Rocks and Minerals*. The authors were Richard D. Dayvult, a geoscientist from Colorado, and H. Steven Hatch of Utah. The material they described was the product of patient collecting carried out between 1998 and 2003, mostly from Mt Ellen. Ten hours of work had been spent on recovering one single specimen, and the new material effectively expanded upon Tidwell and Medlyn's diagnoses. In addition to the short shoots the article also covered associated fossils, including seeds and a probable fern.

Personal communications with Dayvault were to prove invaluable in the identification of plant and reptile bed material. Many specimens were provisionally identified on the basis of photographs, his experienced eye being invaluable in interpreting material otherwise unfamiliar in the UK. Some comparisons between plant and reptile bed specimens and illustrations by Chandler that had previously been overlooked now became apparent.

One 20mm short shoot specimen in particular from the plant and reptile bed (NHM V.65128) identified by Dayvault provides a good match in terms of morphology and size with Chandler's type specimen of *Carpolithus* (now *Behuninia*) *provoensis* from Utah, and

---

246 Tidwell (1998) cited Dayvault and Hatch (2003), 241.

sectioning of this specimen at the NHM revealed a similar internal structure of central pith and secondary xylem to that later described by Tidwell and Medlyn. Fragmentary fossils demonstrating similar structure have been collected by the author from both the plant and reptile bed and the Cherty Freshwater Member of the Purbeck Limestone Group at Durlston Bay,[247] but what had been described by Tidwell and Medlyn as a species of detached shoot could now be seen in Dayvault and Hatch's illustrations in positions of growth, attached to long shoot axes or stalks up to 40cm long. A 32.5cm long stalk from Mt Ellen had four opposite pairs of shoots attached, and a 21cm long shoot from the same site showed a whorled arrangement of short shoots and a change in the dominant growth direction.[248] Various stages of growth, differences in bulbous swelling and in elongation, and attachment surfaces at the base of shoots resulting from possible abscission or shedding had been illustrated to a limited degree by Tidwell and Medlyn,[249] but now Dayvault and Hatch had access to specimens demonstrating a far larger range of such features. Some of these detached shoots reached lengths of up to 12cm, elongating into new stalks or long shoots from which new buds could develop. Owing to their shape they were described as 'typical "bottle-stopper" *provoensis*'. *Behuninia provoensis* short shoots were also illustrated attached to a more robust piece of wood, in the form of a 26cm long branch with two sub-branches.[250] As Dayvault and Hatch described their specimens they found difficulty in making specific identifications to the two species *B. bassii* and *B. scottii* described by Tidwell and Medlyn, stating that 'Although some of the fossils we collected may belong to one of these species, we did not attempt their identification

---

247 Needham (2007), Figs. 32 and 33.
248 Dayvault and Hatch (2003), Figs. 13 and 15.
249 Tidwell and Medlyn (1992), Plate V.
250 Dayvault and Hatch (2003), Fig. 17.

Plant fossils from the Brushy Basin Member of the Henry Mountains of south east Utah attributable to the species Behuninia provoensis. Such finds, some with several shoots attached to stalks over 30 centimetres long, have a comparable growth pattern to plant and reptile bed wood illustrated in the previous chapter.
Specimens: Author's collection, collected Richard Dayvault. Photo: Author.

to this specificity.'[251] Wood specimens demonstrating long shoot-short shoot structure not identifiable to species level were referred to as *Behuninia* sp..[252]

Many of the Utah specimens of *B. provoensis* cannot be equalled in terms of their superb articulation by fossils collected by the author from the plant and reptile bed, but there is nonetheless some very good comparative material. Two specimens of what is provisionally

---

251 Dayvault and Hatch (2003), 241.
252 Dayvault and Hatch (2003), eg. Fig.18.

interpreted as *Behuninia* sp. wood, 25.5cm and 21.5cm long respectively and illustrated in Chapter Two, demonstrate the typical long shoot-short shoot organisation of nodes bearing probable short shoot truncations and internodes formed by long shoot elongation. One shoot from just below the plant and reptile bed, exceptional in that it was collected from a limestone off-cut taken from the top of the Tisbury Member (NHM V.65108), demonstrates a well-developed attachment and a collar of secondary stem xylem covering the base of the shoot, features found in many Utah specimens. The plant and reptile bed has also produced some very delicate specimens, such as a 40mm shoot with a 32mm attachment, and a 19mm shoot with a slight upward curve to a pointed apex. The former specimen is of note for its parallel sides and truncated squared-off end, a feature of some plant and reptile bed specimens not obviously comparable with illustrated American material. The bulbous swelling seen on several American specimens tends to be either absent or situated only at the very base on Wiltshire specimens. Many of the Wiltshire specimens have the appearance of relatively thin, tapering thorns, and could have evolved this morphology as a defensive function against herbivores – indeed, a specimen from Utah later allocated to the species *Behuninia provoensis* was originally described by Chandler as a thorn. There is a distinct possibility, or even a probability, that more than one plant species bore short shoots of the *Behuninia provoensis* type, and plenty of scope for a detailed comparative study of American and English material. What is beyond dispute is the degree of similarity between some specimens, indicating that at least one type of tree or a closely related group of trees bearing *Behuninia provoensis*-type short shoots was common to southern England and the western USA towards the end of the Jurassic period.

If *Behuninia provoensis* can be seen as a morphotaxon rather than a specific plant species, the same is equally true of *Behuninia joannei*, Chandler's cycadophyte seed species redefined by Tidwell and Medlyn as a short shoot. The basic concept of oppositely arranged shoots on stalks with a central pith and secondary xylem covers a broad range of specimens, of which the most spectacular to have been illustrated was collected by Dayvault from the Yellow Cat area north of Moab, Utah. In a species in which most specimens take the form of detached pairs of shoots, the Moab specimen has six pairs of opposing shoots and a single terminal shoot attached to a 29cm long axis.[253] The shoots are rounded and bud-like in a specimen whose articulation suggests no great abrasion before preservation, whereas other specimens such as those illustrated by Tidwell and Medlyn[254] tend to have pointed apices. Dayvault and Hatch's and Chandler's illustrations of detached pairs of shoots range between these two forms.

Specimens from the plant and reptile bed comparable to American *Behuninia joannei* specimens are uncommon and, although some other material shows exceptionally fine preservation, tend to appear worn. Some of these fossils are ambiguous, presenting the bi-lobed appearance of a pair of shoots from one side and the undivided appearance of a single terminal shoot or lignotuber from the opposite side. The largest such specimen lies at the end of a 31mm long axis.[255] A specimen from the plant and reptile bed with a single lobed shoot on a 15mm long axis[256] is comparable with specimens of *B. joannei* described by Dayvault and Hatch. In an unusual specimen from the base of the plant and reptile bed, a 35mm long specimen of *B. provoensis* and a 17mm long specimen with some typical *B. joannei* characteristics are embedded in a single piece of silt.

Readily identifiable specimens of the probable short shoot species *Steinerocaulis radiatus* are moderately frequent in the plant and reptile bed. The distinctive shield-shape of small specimens is

---

253 Dayvault and Hatch (2003), Fig.22.
254 Tidwell and Medlyn (1992), Plate 1.
255 Needham (2007), Fig.27.
256 Needham (2007), Fig.28.

* *Forests of the Dinosaurs* *

Short shoots from the plant and reptile bed. Left, viewed upside down with attachment to the left, is a Behuninia provoensis type shoot. The specimen to the right is a possible pair of Behuninia joannei shoots. The block of matrix, with its rare juxtaposition of examples of the two morphotaxa, could only be retrieved successfully following in situ PVA consolidation of the silt. Specimen: Author's collection. Photo: Author.

identical to that described by Chandler and by Tidwell and Medlyn. Through the study of large numbers of American specimens, Dayvault and Hatch considered the growth of these possible budding shoots into larger structures found in the Brushy Basin Member of the Morrison Formation. Such fossils are comparable to Argentinian specimens from the Cerro Cuadrado Petrified Forest, described as seedlings and aerial lignotubers. Lignotubers are woody structures that grow on some of today's trees including *Ginkgo biloba*. As asexual reproductive structures they can grow into clones of the parent tree when detached from it. As has been described above, occasional plant and reptile bed specimens with the appearance of *B. joannei* short shoots when viewed from one side have the appearance of lignotubers when viewed from the opposite side. Such specimens belong to a growing body of evidence indicating a mix of common and differentiating features between related morphotaxa from geographically diverse sites – another indication of the need for a detailed study overviewing the various morphologies.

Occasional plant and reptile bed specimens appear all but identical to some larger *S. radiatus* specimens described from the Morrison Formation by Dayvault and Hatch and lignotubers described from the Cerro Cuadrado Petrified Forest of Argentinian Patagonia. The best preserved of these, a largely uncompressed lignotuber with a diameter of 14mm and a truncated stalk (NHM V.65123),[257] is remarkably similar to a much larger lignotuber from the Cerro Cuadrado with a diameter of around 60mm.[258] When the Wiltshire fossil was first compared with a photo of the Argentinian specimen, it produced for the author one of those rare moments of discovery in fossil collecting – two small fossils that provide evidence of a former link between the forests of two great continents. It is the building up of this larger picture that places the Tithonian vegetation of southern England into a broad global context, offering glimpses of common features across worldwide Upper Jurassic forests, not through the usual trunks and branches of great trees, but through the detailed structures they produced.

Other specimens from the plant and reptile bed have what appear to be two shoots or lignotubers arranged in sequence along the axis, one terminal and the other forming a possible node. A lignotuber from the Cerro Cuadrado with two swellings has been illustrated,[259] but these have not separated from each other. The plant and reptile bed form, found in varying sizes and states of compression, appears to represent a morphotaxon distinctive to the English flora. A number of further specimens are not easy to interpret, and could represent seedlings or clones growing from lignotubers. There are many difficulties involved in the interpretation of some petrified plant fossils, and many Cerro Cuadrado specimens originally described as

---

257 Needham (2007), Fig.36.
258 Stockey (2002), Fig.1.
259 Stockey (2002), 168.

seedlings have been reinterpreted as shoots and aerial lignotubers.[260] The outlandish appearance of a few plant and reptile bed specimens would make them the most unusual additions to any fossil collection.

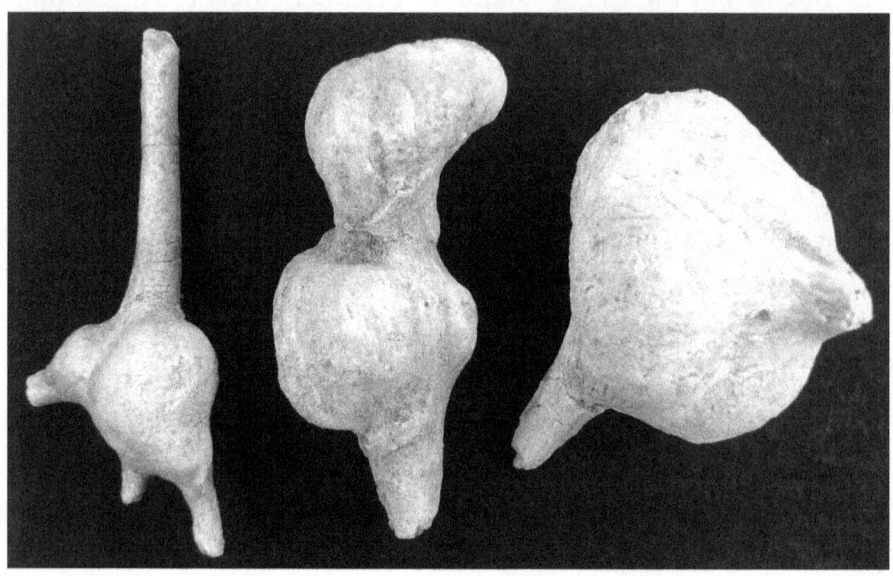

Unusual plant specimens from the plant and reptile bed. The specimen to the left is a possible seedling or atypical shoot, the middle specimen is an undescribed morphotaxon with two short shoots or lignotubers in sequence, and the specimen to the right is a possible lignotuber with small truncations (not visible in photo) that could mark root growth. Specimens: Author's collection. Photo: Author.

The double-chambered seed capsule from the plant and reptile bed described in the preceding chapter, and interpreted as possibly containing *Carpolithes acinus* seeds, was at first compared with Chandler's *Jensensispermum redmondi*.[261] Chandler devoted three pages of illustrations to this species,[262] but it was only three illustrations of a single specimen 'apparently referable to this species' that allowed

260 Stockey (2002), 171.
261 Needham (2007), 6–7.
262 Chandler (1966), Plates 5–7.

any substantial comparison to be made. If this is a case of closely related transatlantic cousins, it seems more likely that Chandler's atypical specimen was not in fact of the same species as the other *J. redmondi* specimens, but the matter remains highly speculative.

There is one type of plant fossil from the plant and reptile bed that, in terms of morphology and abundance, provides an unequivocal common link with both the Purbeck Fossil Forest as represented by the Charophyte Chert of Portesham Quarry and the short shoot locations of the Brushy Basin Member of the Morrison Formation. This is the already familiar seed or cone scale species *Carpolithes westi* described by Brown and Bugg in 1975. Over sixty specimens have been collected from the plant and reptile bed since 2002, making this the commonest plant fossil to be found there. It was also the most abundant species in the Portesham Quarry material. Sizes vary, but average length is around 7 to 8mm. In the Henry Mountains of Utah, at a site on Mt Ellen, similar fossils 'practically litter the surface of certain areas'.[263] Many specimens have a flattened base and appear less abraded than most plant and reptile bed specimens, but others are virtually indistinguishable. They are recognised in the USA as cone scales and examples from the Morrison Formation at the Natural History Museum in London are labelled *Araucaria* sp.. Dayvault and Hatch (2003) note that the Utah specimens show bi-fold symmetry and that some are split to reveal a seed compartment. Small seeds from 0.5 to 2mm in diameter are seen as probably coming from these seed scales, and one seed in this range from the plant and reptile bed could be a specimen of similar origin. 'A tiny shrivelled body of unknown nature' found in sectioned specimens of *Carpolithes westi* from Portesham Quarry[264] could no doubt also be a similar seed. If the specimens from Dorset, Wiltshire and Utah are all cone scales, they

---

263 Dayvault and Hatch (2003), Plate 31.
264 Brown and Bugg (1975), 434.

appear to come from a cone species that readily disintegrates, although it is of note that when Brown and Bugg considered the possibility of *C. westi* specimens being araucarian cone scales, they also saw them as possibly coming from their new cone species *Araucarites sizerae*, and in turn saw this species as most similar to *Araucaria mirabilis* from the Cerro Cuadrado – an interesting if somewhat tenuous chain of connections. Complete identifiable cones are unfortunately absent from material collected from the plant and reptile bed, although the Brushy Basin Member does yield cones which, excepting Chandler's *Sequoia* sp. and *Hillistrobus axelrodi*, are mostly undescribed.

Comparisons between the plant and reptile bed flora and that of the short shoot localities of the western USA thus reveal some remarkable common elements in two morphogenera of short shoots and in one type of seed or cone scale. With a common seed flora occurring in the Purbeck Fossil Forest horizon at Portesham Quarry and the plant and reptile bed flora of Wiltshire, it is worth looking at similarities between the Purbeck and Morrison plant assemblages.

Firstly the preserved material from both floras is no doubt very incomplete, and some of the comparisons are very general – horsetails from Montana and Dorset, ferns from Montana, Wyoming and Utah and from Purbeck miospore assemblages, and *Brachyphyllum* shoots and twigs from Utah and Dorset. *Brachyphyllum* is a morphogenus associated with probable araucarian remains in Jurassic rocks of Yorkshire and with cheirolepidiaceous remains in Cretaceous rocks of Europe,[265] both conifer families with wood morphogenera identified from the Purbeck Fossil Forest. Fossil wood from the Morrison Formation of Wyoming has been referred to *Araucarioxylon*,[266] but as has already been seen the problems involved with Mesozoic wood identification and classification are complex. The tendency

---

265 Taylor, Taylor and Krings (2009), 844–5
266 Tidwell (1990), 4–5.

The attribution of cycadophyte trunks to species level, such as this probable bennettite trunk from Wiltshire's Purbeck Fossil Forest, is not easily made, and some twenty described species from the Brushy Basin member of the western USA were later reassigned to a single species. The £2 coin used for scale has a diameter of 28mm. Specimen: Author's collection. Photo: Author.

to associate these trees with modern forms is widespread, and yet Tidwell has warned that they were evolving transitional forms that combined ancestral and modern characteristics, and that 'if classification of these species is based only on primitive characters, all would be diagnosed as araucarian or, probably more accurately, cordaitalean conifers'.[267] The short shoot genera *Behuninia* and *Steinerocaulis* have so far revealed nothing to help in conifer classification, having been provisionally referred to an unknown family of Coniferales, although anatomically most resembling among extant families the Taxodiaceae.[268]

The cycadophytes described from the Purbeck Fossil Forest have already been covered in Chapter One, where it has been seen that three bennettite species of the genus *Cycadeoidea* were named during the nineteenth century. Differentiation of species on the basis of trunks is not as straightforward as differentiation on the basis of leaves, which may in part explain the much larger species diversity from the Middle Jurassic rocks of Yorkshire where most fossils are compressions. In

---

267   Tidwell (1990), 4.
268   Tidwell and Medlyn (1992), 237.

1900 the American palaeobotanist Lester F. Ward described twenty species of bennettite – he used the name cycadeoid – from the Brushy Basin Member attributed to a new proposed genus *Cycadella*. However, *Cycadella* and most of its species disappeared in 1960 when 'these twenty species were subsequently reassigned to *Cycadeoidea* as the single species *C. wyomingensis*'[269] by T. Delevoryas. Cycadophyte fossils from the Brushy Basin Member of the Morrison Formation of south-eastern Utah have since been further described and illustrated in detail. Apart from the palaeobotanical significance of the trunk specimens they are preserved in beautifully coloured minerals – arguably 'the most beautifully colored cycads in the world'.[270] These illustrated Brushy Basin specimens are not assigned to species level identifications, and the comparison between the Morrison Formation flora and the Purbeck Fossil Forest flora depends here on a basic comparison of gross morphology rather than on detailed species lists.

Once again, the fragmentary nature of the plant fossil record is clear to see. For all but identical specimens to occur in a single plant bed in southern England and across limited strata in four states of the western USA and, in one case, in a small area of Patagonia, clearly indicates that plants growing across a large area of the world disappeared with very little trace left behind them. Even more of a mystery is that they appear to have been preserved by a comparable process of silicification under very different conditions in very different types of deposits. What these conditions were is thus a matter of some interest, as is the ancient climate or palaeoclimate.

The south of England and the western USA were considerably closer to each other in Upper Jurassic times than they are now, with the opening up of the Atlantic in its early stages. The geography of Jurassic England was dominated by global sea-level changes and 'crustal

---

269 Tidwell (1990), 1.
270 Dayvault and Hatch (2005).

extensions related to rifting in the North Atlantic'.[271] The published discovery in 2006 of the dinosaur *Stegosaurus* from uppermost Kimmeridgian – lowermost Tithonian strata in Portugal, previously known only from similarly dated Morrison Formation strata in the USA, reinforced 'previous hypotheses of a close relationship between these two areas during the Late Jurassic'.[272] Evidence of faunal and floral contacts between North Atlantic landmasses during this period are supported both by this find and by geotectonic evidence indicating a 'high probability of an episodic corridor between the Newfoundland and Iberian landmasses'.[273] On the large scale, therefore, the floral similarities between the plant and reptile bed flora and that of the Morrison Formation are readily explicable.

It has been seen how a significant marine regression at the end of the Jurassic period led to the emergence of Purbeck conditions in southern England. In the western USA the Morrison Formation began with marine deposition and, as in southern England, a regression occurred leading to the formation of coastal mudflats and increasingly freshwater conditions. By the time of the Brushy Basin Member deposition the rivers flowed into a landlocked body of increasingly saline water named Lake T'oo'dichi', with volcanically active mountains to the west and ancestral Rocky Mountains to the east. Within the proposed semi-arid climate, an extensive plant assemblage grew along shallow river valleys.[274]

Conditions in the Morrison and Purbeck areas of deposition were thus much less different from each other than they are now, even if current contrasts between temperate and fertile south-central England and continental desert conditions in Utah make it difficult

---

271 Rawson (1992), 107.
272 Escaso et al. (2007), 367.
273 Escaso et al. (2007), 367.
274 The National Park Service www.nps.gov/archive/dino/morrison.htm (visited 2009).

to conceive of a long-gone shared environment. Purbeck altitude was barely above sea level, terrestrial conditions were associated with adjacent freshwater and brackish to hypersaline conditions, and deposition was mainly low energy with a high proportion of limestones. Conditions during deposition of the Brushy Basin Member were admittedly more variable, with low and high energy deposits ranging from limestones, volcanic ash-derived sediments, mudstones and siltstones to sandstones and conglomerates, but altitude 'was probably less than 300 metres above sea-level',[275] and 'The sedimentary environment was a broad, flat flood-plain with meandering streams and numerous shallow lakes or playas that could have become alkaline or saline during drier times.'[276] Bearing in mind that the Morrison Formation and plant and reptile bed short shoots were transported from other locations to their final resting places, both could well thus have come from environments that were similar in several key aspects such as flat lowland topography and variable salinity of adjacent standing water.

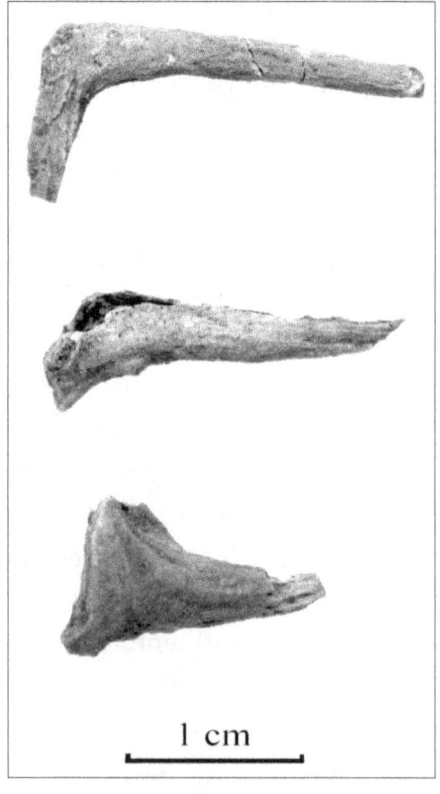

These short shoots from the plant and reptile bed with Behuninia provoensis-type attachments were transported from other localities to their final resting place, but the upper specimen shows far more sign of wear and tear than the middle specimen. These examples indicate the possible thorn function of some short shoots. Specimens: Author's collection. Photo: Steve Clifford.

275   James A. Jensen in Chandler (1966), 140.
276   Dayvault and Hatch (2003), 235.

Climatic conditions could equally have been comparable taking into account relative latitudes and altitudes.

The divergence of opinion over the Purbeck Fossil Forest conditions – a jungle or a luxuriant coastal swamp as against the more likely semi-arid scenario, possibly with marginal conditions for tree growth – has to some extent been reflected in interpretations of the Brushy Basin Member forests. The short shoots have been interpreted as indicating a very different climate from Francis's Purbeck Fossil Forest: 'Thus, the lack of definite growth rings in these shoots and the axes to which they are attached suggest an equable climate without any major changes that would appreciably alter the production of secondary tissues.'[277] Growth rings on *Protocupressinoxylon purbeckensis* wood, to recap, have been interpreted as indicating a seasonal, highly variable climate with prolonged periods of drought. Sectioned wood found in association with short shoots in the plant and reptile bed shows growth rings similar to those of *P. purbeckensis*, and contradictory climatic indicators thus occur together. Other interpretations of the Morrison climate are more in accord with Francis's view of the Purbeck climate. One interpretation 'proposed the climate in the southern part of the Morrison Formation as being close to arid with extended times of dryness interrupted by only short periods of moisture',[278] a view in line with Vakhrameev's proposed widespread Upper Jurassic climate aridisation. It should be noted here that the Morrison Formation covers an extensive geographical area with far more environmental variation than was likely to have been found within the area of the Purbeck Limestone Group deposition.

It is a matter of interest that the fossil plants represented by the short shoots occur abundantly, especially in the western USA, but within a relatively limited time band. The plant and reptile bed

---

277 Tidwell and Medlyn (1992), 237.
278 Peterson and Turner-Peterson (1987) cited Tidwell and Medlyn (1992), 237.

lies several metres at least below the Jurassic-Cretaceous junction, which in turn lies at an indeterminate point in the lower half of the Purbeck Limestone Group. As that junction is dated at 145.5mya and the beginning of the Tithonian at 150.8mya, an estimated age of 146 to 149 million years would seem reasonable. Dating within the Morrison Formation is more specific, as beds of volcanic bentonitic ash allow for radioactive dating. 'The age of the upper Brushy Basin Member of the Morrison Formation (148-150 Ma) is known from ash beds dated from several sections across eastern Utah.'[279] In geological terms, recorded finds of *Behuninia* and *Steinerocaulis* thus appear and disappear from the fossil record relatively rapidly, while other plant morphogenera show continuity from epoch to epoch.

Preservation of the American short shoots is associated with those same volcanic ash deposits that permit radioactive dating. 'It is no coincidence that petrified wood is commonly associated with volcanic ash deposits or in rocks that resulted from ash that has been mobilized and redeposited by water.'[280] In the case of the remarkable Cerro Cuadrado Petrified Forest of Patagonia, the trees were silicified following burial by volcanic ash, having been 'overwhelmed in an outburst of volcanic activity'.[281] The proposed silicification of the plant and reptile bed material through downward circulation of groundwater as proposed by Astin (1987) appears to be a very different process, but with a coincidentally very similar outcome. It is worth considering the possibility that soluble volcanic ashes played a part, as they would provide a ready source of mobile silica leading to potential plant silicification and chert formation. Some clasts or lumps of sediment within the Tisbury Member limestones originate from atypical beds that have since been eroded away, and could provide some evidence for or against this idea.

279 Kowallis, Britt, Greenhalgh and Sprinkel (2007), 75.
280 Daniels and Dayvault (2006), 179.
281 Calder (1953), 101.

Whatever the conditions were actually like, the vegetation supported a wide range of animals in both southern England and the western USA. The faunas of the Purbeck Limestone Group and the plant and reptile bed have already been examined, and a brief transatlantic comparison is now made with the Morrison Formation of the western USA, with a closer look at some of the sauropods. The vertebrate assemblage from this formation is extensive and mostly found in the Brushy Basin Member. Some creatures have been identified on the basis of numerous skeletons in 'dinosaur graveyards', others on the basis of nothing more than the occasional tooth.

Of the smaller vertebrates, frogs are common to the plant and reptile bed subject to formal identification, to the Purbeck of Wiltshire (tentatively) and Dorset and to the Morrison, where more complete material has been found – in February 2008 it was reported that 'Two slabs of Jurassic frog fossils have returned to Dinosaur National Monument after more than a decade at the Carnegie Museum of Natural History.'[282] Salamandroids are recorded from Dorset and from the Dinosaur National Monument. Two turtle genera, *Glyptops* and *Dorsetochelys*, are common to the Purbeck and the Morrison. One

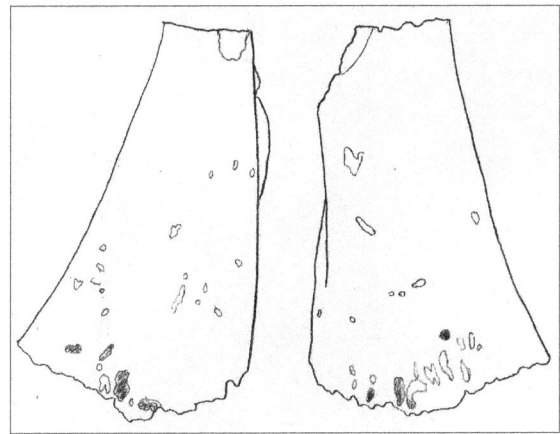

Two line drawings of a small bone from the plant and reptile bed, approximately 4mm long and provisionally identified at the NHM as 'possible frog.' Such fossils are rare from the Jurassic-Cretaceous transition in England, although unconfirmed frog remains from the Purbeck Limestone Group of Swindon were reported as long ago as 1876. Specimen: Author's collection. Drawings: Author, based upon photos by Paul Barrett.

282 The National Park Service www.nps.gov/dino/parknews/frog-fossils-return-to-monument.htm (site visited 17 February 2009)

of three Morrison genera of spenodont lizard, *Opisthias*, is similar to a genus from the Purbeck of Dorset.[283] Several sqamate lizards are found in the Purbeck of Dorset and the Morrison, including the genus *Dorsetisaurus*. Undescribed lizards have been found in the plant and reptile bed.

Of six genera of crocodilians occuring in the Morrison Formation, the greatest distribution and species diversity is found in *Goniopholis*, a genus also recorded from the Purbeck and the plant and reptile bed. Pterosaur remains are rare in the Morrison as well as in the Purbeck, and not yet described from the plant and reptile bed. Only the broadest of comparisons can be made, with pterodactyloids reported from the plant and reptile bed and three pterodactyloid genera from the Morrison. Rhamphorhynchoids are represented in the Purbeck by possible teeth from Sunnydown Farm in Dorset and in the Morrison by two genera.

When it comes to dinosaurs, southern England comes a poor second to the Morrison Formation in terms of the fossils it yields. In the nineteenth century early geologists such as Mantell, Buckland and Owen spent many years painstakingly assembling teeth, bones and fragments of skeletons, making many mistakes along the way and identifying only a small number of genera. The opening up of the western USA in the second half of the nineteenth century also opened up a new chapter in vertebrate palaeontology, as dinosaur skeletons were uncovered in large numbers. In the early twentieth century what is now the Dinosaur National Monument was discovered by American palaeontologist Earl Douglass in the Brushy Basin Member of the Morrison Formation along the Utah/Colorado border. Douglass supervised the removal of countless tons of bones and bone-yielding rock between 1909 and 1924, and there remains on display at the Quarry Visitor Center a rock face containing some 1,500

---

283 Evans and Searle (2002), 145.

bones. Numerous lesser sites range from quarries yielding articulated skeletons down to the Fremont site in Utah, at which dinosaur bones as well as the short shoots described by Chandler as cycadophyte seeds in 1966 were found.

Of the sauropods, the most abundant Morrison genus is *Camarasaurus* – the number of skeletons at Dinosaur National Monument has led to it being considered a possible herd animal. *Camarasaurus* belongs to the sauropod clade Macronaria – a clade is an evolutionary branch – which includes the Family Camarasauridae at its base and the Titanosauriformes, which as the name suggests can be quite large. The Family Brachiosauridae lies at the base of the Titanosauriformes branch, and the first brachiosaur ever described was the rare Morrison *Brachiosaurus altithorax*. *Pelorosaurus* and cf. *Pelorosaurus* teeth from the Purbeck of Buckinghamshire and the plant and reptile bed respectively have been seen as possible brachiosaurids. The camarasaurid record from the UK and from Europe as a whole is somewhat tenuous. According to vertebrate palaeontologist Darren Naish, the Barremian Age dinosaur assemblage from the Lower Cretaceous Wessex Formation of southern England did include 'a possible camarasaurid (*Chondrosteosaurus*)'. This attribution was later revoked, as according to Naish 'the characters supposedly indicating camarasaurid status for *Chondrosteosaurus* are rubbish'.[284] Articulated dinosaur remains with several vertebrae but no skull from the Upper Kimmeridgian to Lower Tithonian of Portugal were attributed firstly to the genus *Apatosaurus*, which incorporates the invalid but more popularly known *Brontosaurus*, but later proposed as a possible *Camarasaurus*. In 1998 the remains were reattributed to a new genus *Lourinhasaurus*, which 'bears similarities to *Camarasaurus*'.[285] Vertebrae can be of limited use in identifying dinosaurs to family

---

284 Darren Naish: Tetrapod Zoology http://darrennaish.blogspot.com/2006/02/lots-of-sauropods-or-just-a-few.html (visited 1 December 2010).

285 Dantas et al. (1998) trans. Harris (2002).

level or below – examples from the Portlandian (Tithonian) of France were 'described and referred to a sauropod dinosaur belonging either to the Camarasauridae or to the Brachiosauridae'.[286]

Sauropod teeth change little between families and are thus as problematic as some bones in sauropod identification, and it was on the basis of teeth that an unpublished 'camarasaurid' had been anonymously reported from Upper Chicksgrove Quarry in the 1980s. This author provisionally identified the two spoon-shaped sauropod tooth crowns collected from the plant and reptile bed in 2010 as *Camarasaurus* sp. after comparison with a replica *Camarasaurus grandis* specimen from the Morrison of Utah, after an examination of images of a *Camarasaurus* skull replica known as ET from the Morrison of Wyoming,[287]

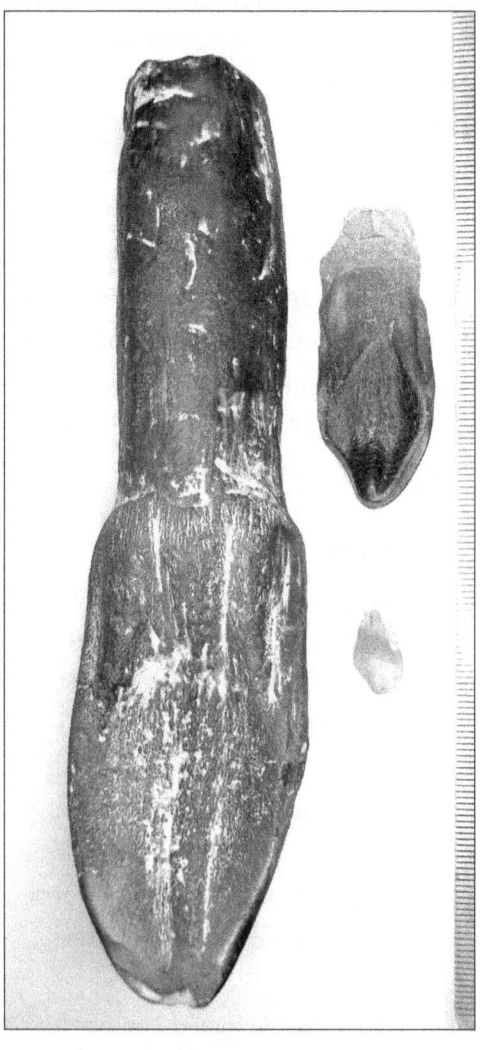

A replica of a Camarasaurus grandis tooth from the Morrison Formation of Utah (left) shares many common features with a 46mm long sauropod tooth crown from the plant and reptile bed (top right). Along with the small tooth crown below, the Wiltshire specimens are technically teeth of 'indeterminate non-diplodocoid non-titanosauriform eusauropods.' Specimens: Author's collection. Photo: Author.

286 Buffetaut (1990).

287 Black Hills Institute www.bhigr.com/store/product.php?productid=305 (visited 1 December 2010).

and after brief research into some diagnostic features. Differentiation between camarasaurid and brachiosaurid teeth presents potential pitfalls. However, the smaller of the two plant and reptile bed teeth, de-enamelled and only 14mm long, appeared to be a laterally expanding crown from the back of the mouth that demonstrated the distinctive spoon-shaped depression of the larger 48mm long tooth, whereas in brachiosaurs the smaller back crowns 'become more compressed, losing the spatulate oval depression'.[288] However, identification by the NHM described the tooth as 'not *Camarasaurus* but similar' and formally named it as an 'indeterminate non-diplodocoid, non-titanosauriform eusauropod'. The less complete specimen from the gastropod micrite described as cf. *Pelorosaurus* could belong to the same indeterminate species.[289]

It would be good to have been able to say more – that on the evidence of a couple of teeth, herds of the well-known and described herbivorous sauropod *Camarasaurus*, up to 18m in length, could have roamed south Wiltshire in Tithonian times. The truth is probably not much different, but the specific identification remains lacking without further material.

Apart from camarasaurs and brachiosaurs in the Macronaria clade, the only other sauropods of relevance here are in the Diplodocoidea clade. Morrison genera in Family Diplodocidae include *Apatosaurus*, *Diplodocus* and *Supersaurus*. These dinosaurs have whiptails as opposed to the shorter, chunkier tails of *Camarasaurus*. *Diplodocus* had 'slender peg-like teeth' as opposed to the 'thick spoon-shaped teeth' of *Camarasaurus* and *Brachiosaurus*.[290] Teeth of a 'diplodocid sauropod' have been reported anonymously from the plant and reptile bed,[291] but the author has found no specimens himself.

288 Lim et al. (2001), 82.
289 Barrett, P. (2010) Pers. comm.
290 Lim et al. (2001), 82.
291 Benton, Hooker and Cook (2005), 54.

Of the theropods, there are representatives of several genera within the Morrison Formation, none of which to the author's knowledge have been positively identified from the Tithonian-Berriasian rocks of the UK. Members of the Carnosauria clade include *Allosaurus* and other allosaurids, a group with which one tooth each from the Purbeck of Dorset and the plant and reptile bed have been compared. Members of the Coelurosauria clade, which incorporates the Maniraptora and hence all present day birds, include among others a troodontid with affinities to the dromaeosaurids from the Morrison Formation, the dromaeosaur *Nuthetes* from the Purbeck of Dorset and Wiltshire and cf. *Nuthetes* from the plant and reptile bed. The record for small theropods appears to be too incomplete for any detailed comparisons to be made.

The situation is a little better with the bird-hipped dinosaurs of Order Ornithischia, but not much. A heterodontosaur from the Morrison was referred to as cf. *Echinodon* but, despite similarities with the Purbeck *Echinodon* from Dorset, is now classified separately as *Fruitadens*. The camptosaurid genus *Camptosaurus* is represented in the Morrison by *C. dispar* and in the Purbeck of Dorset by *C. hoggii*, originally described as *Iguanodon hoggii*. Of the armoured thyreophorans, *Stegosaurus* is found at many Morrison sites and ankylosaurs are also present, but no stegosaurid fossils have been confirmed from the Purbeck or plant and reptile bed and possible ankylosaur fossils only number two, both having been described under the heading Thyreophora *incertae cedis*.[292] The presence of nodosaurs or stegosaurs in the plant and reptile bed is noted in the 1983 anonymous report but cannot be confirmed from the author's own collecting.

The remains of small mammals occur in the Purbeck of Dorset and Wiltshire, the plant and reptile bed according to the

---

292 Norman and Barrett (2002), 183.

1983 report, and the Morrison Formation. The extensive Morrison assemblage includes representatives of groups occurring in the Purbeck of Dorset and Wiltshire and the plant and reptile bed,[293] such as docodonts, multituberculates, triconodonts and dryolestids. There are few common genera – for example, the only common dryolestid genus (from the Morrison and Purbeck of Dorset) is *Amblotherium*, described on the basis of lower teeth[294] – but the diversity indicates that similar ecological niches were filled by a range of probable seed-eaters, insectivores and carnivores. These early mammals, or mammaliaforms, are one of the main areas of Mesozoic microvertebrate research, only touched on here in the briefest of ways.

If exact species matches and more complete faunal lists cannot be made owing to the sparseness of the material, there is little doubt that the vertebrate animal life in what is now the western USA was not hugely different in Morrison times from life in what is now south-central England, although there were no doubt many differences in the detail. When this is

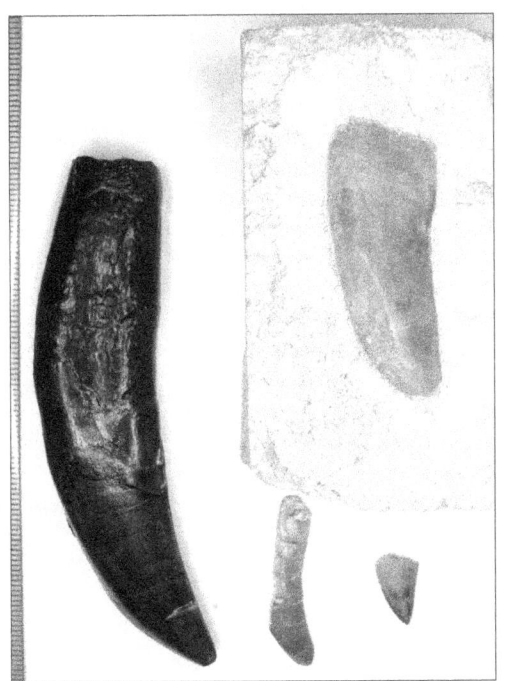

A replica of an Allosaurus tooth from the Morrison Formation of Utah (left), a worn, indeterminate allosauroid tooth crown from the gastropod micrite (top right), a partial tooth crown from the plant and reptile bed (bottom centre) and a broken tooth crown from the gastropod micrite (bottom right). Specimens: Author's collection. Photo: Author.

293 Benton, Hooker and Cook (2005), 55.
294 Trevor Dykes *Mesozoic Eucynodonts* http://home.arco.de/ktdykes/meseucaz.htm (visited 3 December 2010).

considered in conjunction with the floral evidence, with probable forests of similar short-shoot bearing trees extending across both areas, a broader perspective of life in these times is obtained. Assemblages from these geological formations no doubt represent more than one environment, but the Upper Jurassic dinosaur world was climatically less varied with a far smaller diversity of plant and animal species than is the case now. The remains that it left behind have so far been examined as transported debris and the partially preserved stumps and trunks of an *in situ* fossil forest, but what would it be like to see an *in situ* fossil forest preserved in more intricate detail? Some at least of the answers to this question remained locked up until very recently in the basal beds of the Purbeck Limestone Group of south-west Wiltshire, but their discovery was to begin with a chance find made in 2008.

5
SECRETS OF WILTSHIRE'S FOSSIL FOREST

WITH ITS PLACE in the history of English geology enhanced through the works of Miss Etheldred Benett, through its ammonites from the Portland Stone Formation, through its nineteenth-century Purbeck Fossil Forest finds from Chilmark Ravine, through its mammal, reptile and other finds from Swindon, through its fish, reptile and insect finds from Teffont Evias, and through its vertebrate and plant assemblage from the plant and reptile bed, it could be said that by 2008 Wiltshire had played a more than respectable part in the reconstruction of the marine, lagoonal and terrestrial environments of the Cretaceous-Jurassic transition – of the Tithonian and Berriasian ages, that is to say – in Great Britain and in Europe, not to mention within a global setting. Fossils that open up new horizons are not easy to come by, and it was thus by a remarkable piece of good fortune that the Vale of Wardour was to produce another world-class assemblage of plant remains between 2008 to 2010, this time from the basal beds of the Purbeck Limestone Group.

It began when the author decided it would be a good idea, if possible, to supplement the plant and reptile bed fossils with some examples of fossil wood from south-west Wiltshire's Purbeck Fossil Forest horizon, and a new exposure offered a potential collecting site

Stromatolite mounds on an eroded limestone platform adjacent to south-west Wiltshire's Fossil Forest horizon. The mounds, in this case formed in shallow lagoonal waters, normally result from bacterial and algal activity, but can be precipitated inorganically. Examples of such fossils date back around 3.5 billion years, and stromatolites still grow today in Australia. Photo: Author.

in 2008. The exposure revealed a range of strata typical of the Portland-Purbeck transition in Dorset, in the form of marine limestones giving way to thin laminated limestones, stromatolites or algal limestone mounds, clays and carbonaceous palaeosols. There appeared to be two or three palaeosol beds, the lower ones thin and discontinuous, the uppermost one thicker with abundant carbonaceous matter and, as is the case with the Great Dirt Bed of the Isle of Portland, containing worn limestone and chert pebbles. These could well correspond to two 'Dirt Beds' identified at Chilmark Ravine in a nineteenth-century section by H.B. Woodward.[295]

---

295 Cited Reid (1903), 21.

So it was that one day in May 2008 the author visited the site with his daughter Isabelle, who after some surface collecting observed a piece of fossil wood projecting from the ground. It was weathered and the exposed surface damaged, much of the detail having been lost to reveal a bleached, weathered core with some iron staining. It barely seemed worth a second look, but with a spade to hand it was easy enough to dig from the ground.

The weathered piece of fossil wood led to a succession of further pieces. Four sections of wood were extracted with relatively little ease, dipping slightly beneath the exposed ground surface and requiring only a small amount of excavation. Beyond here the specimens began to reveal far better preservation, especially of surface detail. The next section was some 68cm across and 26cm long, and around this time it was noticed that the truncated top of a silicified tree stump lay a few feet beyond the first piece found. This led to the reasonable conjecture that the wood sections were a fallen tree trunk lying by the stump from which it had toppled. The sections were heavily compressed, the 68cm wide specimen having a depth of only 15cm, although later examination revealed that some of the difference was due to incomplete silicification.

With the expectation of an at least respectable length of Purbeck tree trunk, probably of the described species *Protocupressinoxylon purbeckensis*, and with permission to extract and remove the find, further work continued during the evenings of the summer of 2008. The two largest and best preserved sections of trunk followed, one being 56cm long, 73cm across and 20cm deep. The main problem at this stage was of size and weight, with the largest sections weighing up to an estimated 75 to 100kg. An increasing amount of overburden had to be removed over each section, consisting of fragmented, weathered stromatolitic limestone and clay, and the sections were firmly embedded in the sticky clayey palaeosol. The underside of the wood was blackened with a thin carbonaceous coating and revealed a distinctive pattern of rows of depressions apparently resulting from incomplete silicification.

Two longitudinally split sections of trunk lying across a branch onto which they had fallen, and over which the trunk had flexed, were followed by another complete section. Beyond that, some 3.5m from the first piece collected, lay a section that had embedded itself deeply into a depression in the palaeosol and was marked by a considerable thickening of the trunk, reaching a maximum of about 60cm across and 29cm in depth. Removing it was to prove a lengthy process, but when it finally gave up the position it had occupied for some 146 million years, the reason for the thickening became apparent – the trunk divided into two at this point.

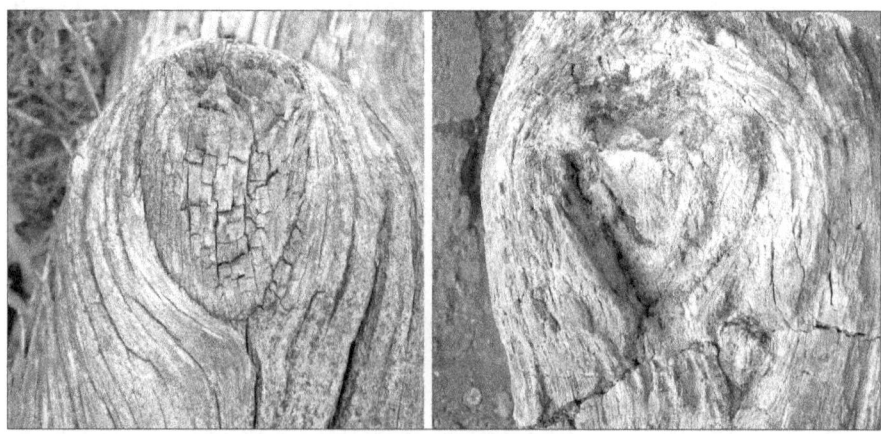

A truncated branch base emerging from a collar of parent branch secondary xylem on a dead mountain pine in the Eastern Pyrenees (left) and a similar feature (right) on section B10/B11 of the 146 million year old fossil tree from the Purbeck Fossil Forest horizon of the Vale of Wardour. Specimen: NHM collections. Photos: Author.

A tree with some branching structure was an unexpected surprise, but practical problems were presenting themselves. It was a lengthy job removing the material from the collecting site, laying it out and cleaning off the caked palaeosol. It was covering an ever larger area of the patio in the author's small garden. Further collecting now required a considerable amount of hand-digging through what was not the easiest of overburden. The next part of the tree proved extremely challenging as it was made up not of neatly formed sections, but of highly fragmentary material in two layers that were at times difficult to sort one from the other. Where the tree divided, one branch had fallen onto the other. It began to look as if reassembly would be difficult to impossible, and a lot of smaller fragments were simply put in bags labelled with the approximate point from which they came. Some 1.5m of this fragmentary material, however, was to prove worth the effort of persevering with. The two branches eventually became clearly distinguishable and removable in complete sections. The upper branch came to an end in a thickened section where, as had happened with the trunk, it appeared to divide, this time petering out rather than demonstrating clear truncated ends. The other branch, seen as probably representing the main trunk, continued for a short distance further until it, too, thickened around 6m from the trunk base, in this case clearly dividing into two truncated ends.

By this stage it was clear that pieces needed to be individually marked for reconstruction, because although they were laid out in order as far as was possible, there was a risk of ending up with an insoluble jigsaw puzzle. Breaks were marked with black enamel paint, using an A prefix for the main trunk, whose sections were marked A1 to A11. The main, lower branch was numbered B1 to B13, and the truncated upper branch C1 to C11. Where sections were fragmented sub-divisions were marked thus: C9; C9'; C9"; C9'''; C9"". Obvious knot positions were noted, of which there were some eight randomly situated along the trunk with diameters mostly in the 2 to 10cm range. There was evidence of one unpreserved branch having had a diameter of 17cm. Nine scattered knots were observed on branch B, including the preserved 2.5 by 4.5cm base of an angled branch. A lack of clearly observable knots on branch C was probably, in part at least, because of poor and incomplete preservation.

By this time the trench from which the tree was being excavated was around 1m deep, and two small branches departed oppositely from the thickened top of branch B at approximate right angles. Although they proved to be no more than 0.6 and 0.7m long, digging them out required the removal of a disproportionate amount of overburden relative to their size. These branches showed no evidence of sub-branching, but had respectively four and five randomly positioned shoots or minor knots with diameters of between 4 and 15mm. In one case a small example of what superficially at least resembled a *Behuninia provoensis*-type short shoot was compressed against the branch.

Direct connections to the tree could not be established for these small branches, but positions in the palaeosol indicated that they were subsidiary branches from the main division at the top of B. Of the two main branches indicated by truncations, no sign could be found of one of them, the truncated end leading to no more than splintered fragments. The other branch could be connected to one side of the truncation, the remainder of the branch base apparently not having been preserved. As had happened to a lesser extent to the fragmented branch bases above the division of the main trunk, it appeared that wood tissue above the swellings or branching nodes had suffered impact damage as the tree fell, and that this had been reflected in subsequent poor silicification.

From the thickened dividing top of branch B, one medium-sized branch leaving at right angles and represented by only two sections was numbered with the prefix E. The connection here was tenuous, however, and the possibility that this was part of a crossed

(opposite page) Wiltshire's fossil tree as reassembled in November 2008. From the apex (foreground): branch H with cycadophyte to fork with small branch I (to right), branch D to junction with small opposite branches F and G and truncated fork, branch B (right) and branch C (left) to fork in main trunk, main trunk A to kneeling figure. Specimen: NHM collections. Photo: Steve Clifford.

※ *Secrets of Wiltshire's Fossil Forest* ※

over branch very real. The two smaller branches described above were prefixed F and G, and the branch that continued on upwards as the main axis was prefixed D.

The apparent structure of the tree so far had been straightforward: a single trunk that forked into two branches that each divided again, with some minor branching points mostly situated at random. Branch D, just under 1.5m long, deflected to one side where what appeared to have been a large, angled branch had broken away, and ended among a jumble of small pieces. Some at least of these turned out to be the base of a subsidiary branch angling away at approximately 45 degrees, which was labelled as branch I, and a continuation of the now slimmer main axis was labelled as branch H. A direct physical connection could not be made to the main branch D from branch I, probably owing to impact damage when the tree fell. That there was such a link seems beyond reasonable doubt but not an absolute scientific certainty.

Branch I was not easy to collect. The excavation was coming on for 1.2m deep, and the previously fragmentary overburden was now interspersed with the occasional large limestone stomatolite. The end of the branch had to be laboriously extracted from beneath one of these, the alternative being a major removal of overburden. In its entirety branch I turned out to be almost 1m long, heavily compressed and erratically preserved, but with features not found anywhere else on the tree.

Some 35cm from its angled base the branch, with a width ranging from 6 to 10cm, came to a swelling or node with the truncated base of a branchlet on the underside. The diameter of this truncation was 3.5cm and the preservation remarkable, revealing what appeared to be surface texture of the branch bark and, in one small area, branchlet bark textured at right angles. Beyond the truncation a body of silicified wood lay compressed beneath the branch, possibly a branchlet remnant but in overall morphology resembling a large

Behuninia-type short shoot or a compressed, distorted lignotuber. A further branchlet base occured a few centimetres further along on the side of the branch, possibly displaced by compression, with two small bud-like features in the upper angle. Some 65cm along the branch a pair of truncated branchlets were located at the opposite sides of a second swelling or node. These features are characteristic of Morrison Formation wood of the type described as *Behuninia* sp. and also share points of comparison with specimens from the plant and reptile bed.

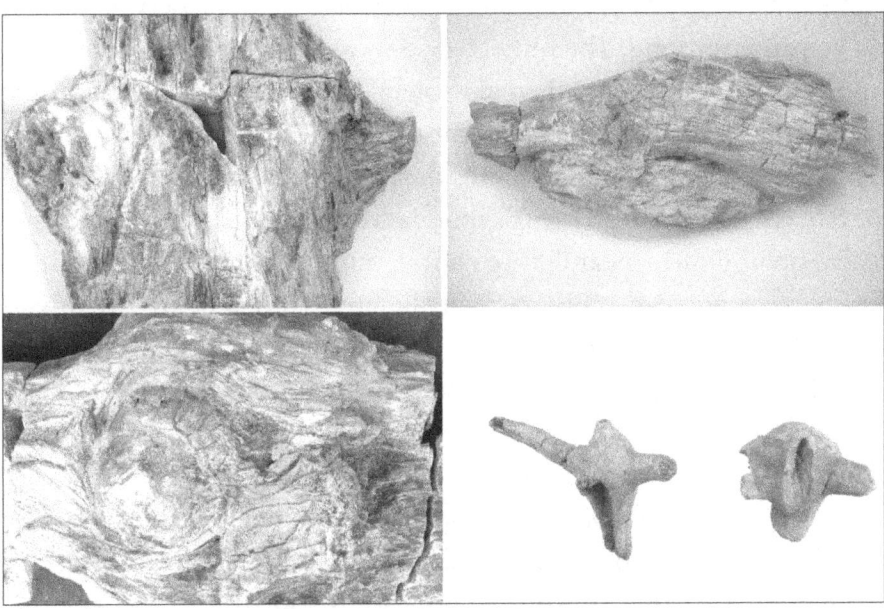

Details of branch I of Wiltshire's fossil tree (top, and bottom left) and of young shoots from the 'plant and reptile bed' (bottom right) from which such a branch type could have originated. The detached piece at top right, probably a branch remnant but possibly a compressed short shoot/lignotuber type structure, originated from the truncated base with bark (bottom left). Specimens: Branch I NHM collections; shoots author's collection. Photos: Author.

By now the fossil tree was spreading beyond the author's patio and onto the lawn, and before continuing with any further collecting it was moved to a lock-up workshop and laid out in sections on the floor. Pieces not yet cleaned were washed with water and a soft brush, smaller fragments where possible were reassembled with PVA glue into more manageable pieces, and the silicified wood was then dried before the labelling was completed and checked. Pieces that could not be fitted into place were put in bags clearly marked with the reference of the nearest break to which they were found.

At the continuation of the main axis of branch D, now labelled as branch H, the fossil wood was about 24cm wide and 7cm deep, and came out of the ground in easily manageable sections between 9 and 29 centimetres long. The only obstacle was the growing overburden, and some evenings were devoted solely to working through this on a trench that was to reach a maximum depth of 1.5m, the work dogged by increasingly hard stromatolites that refused to break up. Along the next 2.1m from H1 to H13 the main axis tapered gently to a width of 19cm on a slightly meandering course, and five randomly positioned knots were mapped. One marked the approximately 4cm diameter base of a branch angling upwards, another the 1.5cm diameter base of a shoot branching at approximately 90 degrees.

Beyond H13 a large roundish projection to the right of the axis was exposed and carefully removed, coming out in four pieces measuring a total of about 23cm across and 8cm deep. The underside was compressed onto two minor branches also apparently originating from branch H. This unexpected bonus was something of an enigma, and various speculative proposals were put forward. The author first considered the possibilities of an aerial lignotuber or a burl, and for some time it remained a mystery, until Paul Kenrick from the NHM saw it and put forward the possibility of a cycadophyte trunk, which later finds confirmed. Its location on the tree is somewhat enigmatic – it either grew there some 10m above ground level as an epiphyte, or plant that grows on another plant, before the tree fell down, which seems scarcely credible, or it grew on the ground before the tree fell down, in which case an explanation needs to be given as to why it is compressed *onto* rather than *under* two small branches, or it grew over the fallen tree and became compressed against it, which would mean that the tree had fallen down before any flooding of the forest by hypersaline lagoon water took place, or it was transported to its final resting place after the tree had fallen.

About 10.5m of tree had now been exposed since discovery of the first piece of trunk, and shortly beyond arrival at the cycadophyte between H15 and H16 there were two knots indicating the departure of a 9cm diameter branch and a 2cm diameter branch. The continuing main axis turned slightly and reduced to an average width of 10cm and depth of 4cm. There was, as it turned out, just under 1m and six more sections to go. A small bud with a diameter of 4mm just beyond H19 was remarkably well preserved, looking like a possible preventitious bud that could have sat dormantly beneath the bark, ready to grow when triggered by a stimulant such as forest fire. A small knot lay just above H20, and H21 was followed by the apex.

Removal of the last three pieces, true to form, took many hours of work and required mechanical assistance as they were overlain by a huge monolithic stromatolite. The total length from trunk base to apex was about 11.5m, and it could be conjectured that 1m at least between the stump and the trunk was missing, owing to a mixture of weathering and ground disturbance. Fragmentary material around the apex suggested that minor shooting took place there, and allowing for upward growth of this a tree with a height in the region of 13 to 15m could be imagined.

In November 2008 the tree, excluding the stump, was reconstructed in the courtyard of a former barn complex and seen by palaeobotanists Paul Kenrick and Peta Hayes of the NHM. That it represented an extraordinarily complete pathway through the branching system of an Upper Jurassic tree was plain to see, and it was offered to the NHM as the most complete fossil tree known to have been excavated from British Mesozoic strata. The offer was accepted, and the move from Wiltshire to museum premises was provisionally planned for the spring of 2009.

The stump meanwhile presented problems of its own. Originally embedded in the ground in two pieces, the smaller of these was excavated and moved to the surface using large crowbars, but the other piece was immovable without mechanical assistance. At a rough estimate the combined weight of the two pieces was 0.75 tonne, and when the larger of these was lifted with the aid of a machine they were both moved on pallets to the site of the lock-up where the remainder of the tree was stored in preparation for transport to the NHM. No serious attempt to examine them was made by the author as they were too heavy to turn, and cleaning was left to the museum, but some points were observed. The smaller piece that had been moved manually had a large root truncation, and some pieces of

Wiltshire's fossil tree as reconstructed in November 2008, with the background blacked out. As collected, the large branch to the right lay on top of the branch continuing upwards from the main trunk. The angles and positions of the small branches are not quite as found in situ, but the essential structure is captured well. Specimen: NHM collections. Photo: Steve Clifford.

probable large roots were later recovered from the vicinity. The truncated centre of the upper surface where the trunk would have continued was preserved in a hard, black, almost flint-like silica, the process of silicification having proceded a stage further than in the rest of the tree.

One of the most interesting features of the stump was a succession of ridges and furrows, exposed where outer, smoother layers of silicified wood were locally absent. In the summer of 2009 the author observed similar, although more widely spaced, ridges around the lower trunk of a dead mountain pine *Pinus mugo* subsp. *uncinata* in the Eastern Pyrenees. The fallen tree was partially decayed and had lost outer layers of wood, and the trunk thickness indicated that it had been of some age when it died. Such features appear to have been a response to compression from the overlying wood.

In the early spring of 2009 the tree, stump and all, was loaded onto a lorry for transport to an NHM storage facility in London. By a fortuitous chance the BBC's Natural History Unit had embarked on producing a series of programmes about the NHM, and the story of the fossil tree was chosen as a subject for filming. The loading of the numerous numbered sections of trunk and branches onto several pallets was thus recorded, carried out by a team that included Paul Kenrick and temporary plant curator Hillary Ketchum from the NHM, the author, daughter Isabelle Needham who had found the tree, and a friend of the author Brian Annetts. Along with the film team, including director, assistant producer, cameraman, sound recordist and liaison officer, it made for quite a send-off.

In May the author and Isabelle went to London to assist Paul Kenrick and an NHM team in reconstruction of the tree on the floor of their storage facility, an event also filmed by the BBC. It was subsequently disassembled and remains in storage, but will hopefully will be put on at least temporary display in the not too distant future. Further filming by the BBC in the area of the discovery site and the digital recreation of a Purbeck lagoon with the tree followed, and in March 2010 edited material was broadcast in the second episode of the six-part BBC2 series *Museum of Life* about the scientific work of the NHM.

Ridges running around the stump of Wiltshire's fossil tree are partially exposed where outer layers of wood are not preserved (left). Similar but more widely spaced ridges can be seen on a fallen mountain pine in the Eastern Pyrenees (right). In both cases the wood has been deformed, probably by compression. Specimen: NHM collections. Photos: Author.

In the Summer 2010 issue of the NHM magazine *Evolve* an article by Paul Kenrick on the fossil tree described in outline the results of initial research.[296] Laser scanning had been employed to 'assemble a highly accurate virtual model of the trunk'. Microscopic examination of cell walls showed the tree to be a conifer probably belonging to the Cheirolepidaceae, and indications were that the tree died following flooding by an adjacent hypersaline lagoon. The branching and rebranching structure of the tree, 'the most complete of its kind in the UK', gave 'new insights into how the tree grew and a more accurate picture of the canopy structure of the forest'. Further details must await the publication of a full scientific paper.

296 Kenrick (2010), 38–9.

The tree had fallen along an approximate north-north-west to south-south-east orientation, and further excavations in its vicinity were to uncover more fossil wood during 2010, most notably two branches lying to the east of the main trunk and probably part of the original tree. Traceable from the edge of disturbed ground about 7m from the base of the trunk, one of these branches extended for approximately 4.7m to a truncated end from which one small short sub-branch curved back in the direction of the tree apex. A probable second sub-branch, not directly connected, appeared to have broken away and continued south eastwards for about 3m, crossing a second large branch some 3.2m long with a forked end that also originated at the edge of the disturbed ground. The angle of the fork was about 60 degrees, and the two resulting sub-branches of quite different sizes. These branches could have been continuations of the truncated upper ends of branches B and C of the tree had they broken away or twisted beneath branch D when the tree fell, but the evidence to verify this one way or the other was absent.

The branches provided some additional information not available from the main tree specimen. On the larger 4.7m branch, with a diameter along the plane of compression ranging from 45cm down to around 30cm, there was a distinct compression ridge on the inside of a curve in the wood similar to the ridges observed on the stump, but otherwise not seen along the branches. At a point where the branch diameter was about 35cm a 15mm long *Behuninia*-like probable short shoot lay compressed against the branch, the position indicating that it had originally grown out at an approximate right angle. Evidence of possible short shoots or bud-like structures on the main tree had been associated with far smaller branches. An eye-shaped knot some 10cm by 12cm contained what appear to be numerous truncated shoots, possibly a response to trauma. The smaller 3.2m forked branch, with the diameter along the plane of compression ranging from 22cm to 17cm, also had a compression ridge on the inside of a curve in the wood.

A 15mm long Behuninia-like bud or short shoot on a 35cm diameter section of conifer branch excavated near the fossil tree. It has been preserved through lateral compression against the branch, and in its original state probably grew out at right angles. This is the largest piece of wood to which the author has found such a structure in direct attachment. Specimen: Author's collection. Photo: Author.

The branches also revealed evidence of wood boring, a feature observed by Paul Kenrick in the main tree. Good preservation of this was found by the author in a length of another branch excavated to the west of the tree. A broken end revealed a number of borings 3 to 4mm in diameter cut across at varying angles. Some were hollow tubes with a light blue mineral lining, others were infilled with silicified material both lighter and darker in colour than the surrounding wood. The infill has the appearance of packed layers of frass, which is the detritus of faecal remains and sawdust left behind by boring larvae. Cracks along the lines of rays in the wood, probably caused by shrinkage, pass from the wood tissue into the borings and have been infilled with silica, indicating that contraction occurred after boring. A section of the same branch broke at one point to reveal part of a lined spherical chamber, possibly a pupation chamber.

The lack of wood boring in Dorset material has been used to support the theory that Purbeck Fossil Forest trees were killed by rising hypersaline lagoonal water, as insects would have been unable to access such submerged or waterlogged wood. If such boring took place after the death of the wood in the Wiltshire material, it suggests either that the wood was already on the ground and bored before water levels rose or that boring took place on dead branches standing above water level. The possibility of activity by other boring invertebrates, possibly aquatic, has to be considered, as trace fossils produced by very different creatures can be notoriously similar.

One advantage of an *in situ* fossil forest is the preservation of roots, which are usually absent or indeterminate in deposits such as the plant and reptile bed. As is the case with other types of wood, their preservation is usually incomplete. Silicified roots from the Purbeck Fossil Forest of Dorset as observed at Chalbury Camp in Dorset were only silicified along the core, 'contained within a continuous lignitic sheath about 3 cm wide'. Roots less than 3cm across were only preserved in lignite, and 'the finer roots have been lost'.[297] Rootlet compressions from the Lower Dirt Bed of the Isle of Portland with diameters of down to 0.6mm were interpreted by comparison with modern trees as probable mycorrhizae.[298] A mycorrhiza is 'a symbiotic association between fungi and the roots of vascular plants'.[299]

A number of specimens collected from the vicinity of the Wiltshire tree were later identified as roots, and where collected *in situ* these were always buried within the palaeosol rather than lying partially exposed on the upper surface. Typical specimens had a

---

297   Francis (1983), 290.
298   Francis (1983), 290.
299   Taylor, Taylor and Krings (2009), 1040.

diameter of about 2cm on the vertical plane and 3cm on the horizontal plane, or a compression ratio of about 1:1.5 – this compared with an average of 1:3 to 1:3.5 for branches exposed on the palaeosol surface. In one case a root was excavated over a distance of 1.5m. Some larger but shorter specimens ranged up to 5.5cm in diameter and a few ranged down to below 1cm. The *in situ* roots were erratically silicified, with some sections not preserved at all and others revealing exceptional surface detail and, in some cases, rootlets ranging from a fraction of a millimetre to about 4mm in diameter running along the outer surface. Most of these rootlets were in the region of 1mm in diameter and at first they were mistaken by the author for climbing plants, as in one notable example three rootlets spread along a main root – which was thought to be a branch at the time – in the manner of a spreading miniature ivy growing up a branch. One specimen with a 4mm diameter axis attached to the surface looked similar in general morphology to a modern host and climber, and a published illustration of several axes of a Permian vine-like climbing fern growing on the surface of a tree fern[300] reinforced this superficial impression.

Formal identification of this material as roots rather than branches was resolved firstly by the sectioning of specimens sent to Paul Kenrick at the NHM,[301] and secondly by the collection of longer roots which clearly meandered through the palaeosol in the manner of a root meandering through the soil. Sectioned specimens exhibited conifer cell structure in the rootlets, although only the larger ones around 4mm in diameter were cut through. They also revealed radially asymmetrical growth around the central axis and the preservation of periderm or outer tissue. In one of the main roots resin canals could be seen passing through the secondary xylem.

300 Rößler (2002), 108.
301 Kenrick (2009) Pers. comm.

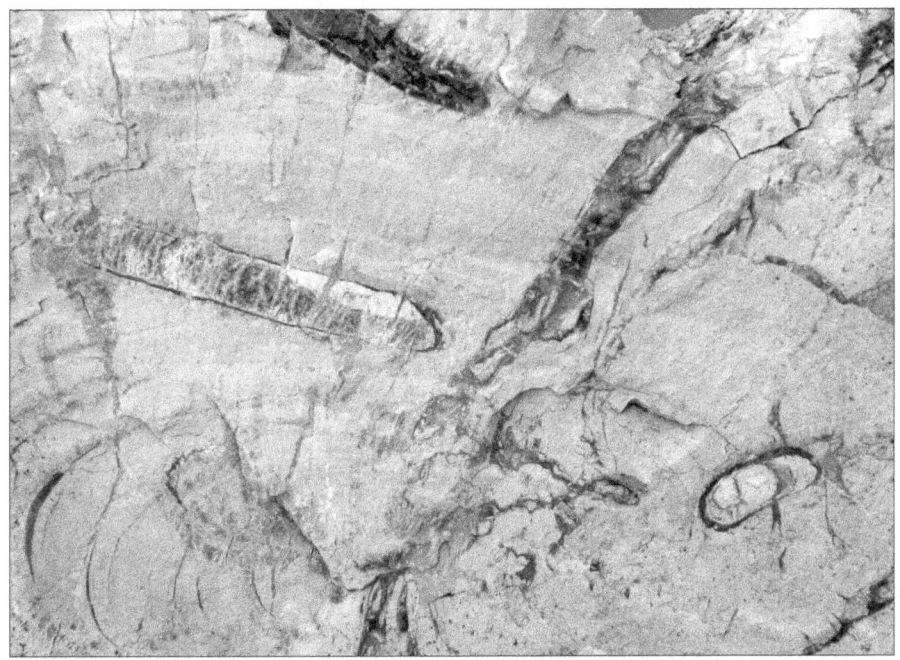

This section of wood from Wiltshire's Fossil Forest horizon has been bored by insects, possibly beetle larvae. The boring to left of centre has been backfilled with frass, a mixture of sawdust and droppings deposited in mounded layers that can clearly be seen. Evidence of boring is absent from conifer wood in the Fossil Forest of Dorset.
Specimen: Author's collection. Photo: Author.

On some roots the surface was slightly crinkled into ridges and furrows in a concertina style, possibly for storage or improved anchorage to the ground. Roots were no more than a few centimetres below the surface of the generally thin palaeosol. One specimen assembled from some surface-collected fragments revealed apparent bark detail in the form of longitudinal irregular surface cracks. Unusual features on this specimen were small node-like branching points and a coating of minute parallel grooves and furrows no more than about 1.25mm across running round the root. These appear to have resulted either from secondary mineralisation or from the mineralisation of algal growth. The possibility of non-organic mineralisation processes being at work here has to be considered, and such processes can be extremely deceptive – examples include Indian stone paving with mineral dendrites interpreted as fossil plants, silicified barite concretions seen on sale in the USA as fossil cycads,[302] and crystalline rock from deposits known to quarrymen as 'beef' in Dorset labelled as fossil wood.

The rootlets raise a number of questions, apart from the obvious one as to how such extraordinarily fine and detailed preservation ever came to take place in stone. Although the larger ones tend to follow a straight course along the outside of the main roots, some of the smaller ones course in the manner of veins and the smallest hair-like examples resemble fine tendrils. A strong clinging mechanism seems to have been at work. Whether the rootlets are of the same species as the roots to which they are attached, or whether they attached to dead roots as saprophytes, or whether they are from a species in a symbiotic or parasitic relationship, remains to be discovered. In two root specimens, apparently healthy looking growth gives way to irregular unhealthy looking wood with the central core of the root, which bears a remarkable resemblance to one of the rootlets, fully exposed along one side. Two root specimens also have transverse structures attached tangentially to surfaces occupied by rootlets.

The roots described above were collected from within some 10 to 12m of the tree stump or from material removed from that area, with the exception of the piece showing the white mineralised surface. The largest piece in cross section, 5.5cm in diameter, came from directly beneath the top part of the 4.7m branch excavated to the east of the tree. This material could thus come from preserved parts of the tree root system.

One last specimen to be mentioned in this root section is a small, knobbly piece of bent wood some 55mm in length by 6mm in diameter. It bears some similarity to specimens from the Morrison

---

302 Dayvault and Hatch (2005), 431.

Formation described as probable segments of fern rhizomes[303] – rhizomes are horizontal underground shoots rather than roots. However, it could equally be of coniferous origin and would require thin sectioning for formal identification.

The discovery of a cycadophyte trunk attached to the upper part of the tree was to be followed by further cycadophyte finds when the excavation of the tree and removal of the stump had been completed. The significance of these plants in the Jurassic vegetation has already been outlined, and there was a sense of history in making new finds from the Wiltshire equivalent of the Dorset Fossil Forest horizon, where the recorded history of bennettites began with Buckland's 1828 paper. Although there seems little doubt that the specimens to be described here are bennettites as opposed to true cycads, in the interests of caution and without formal identification they are referred to under the umbrella title of cycadophytes. The typical cone structures of the Dorset bennettites are not readily identifiable in the Wiltshire specimens, and although there is evidence of such cones it would require sectioning or CT scanning to confirm this.

Many cycadophyte trunks, including the Purbeck *Cycadeoidea*, were sub-spherical to pineapple-shaped structures with fronds growing from the top.[304] They often underwent subsequent compression before fossilisation – if they were preserved in an upright position the height was compressed, and if they had fallen onto their sides they retained their height but became oval in cross-section. The internal structure consisted of central pith surrounded by the vascular tissues of xylem and phloem, all encased in a protective armour of ramenta – a ramentum is a 'hairylike scaly covering, or armor'[305] – that formed diamond-shaped leaf bases from which the petioles or stalks of the fronds emerged. These fronds or trophophylls were the main leaves,

---

303 Dayvault and Hatch (2003), 243.
304 Thomas and Batten (2001b), 142.
305 Taylor, Taylor and Krings (2009), 1044.

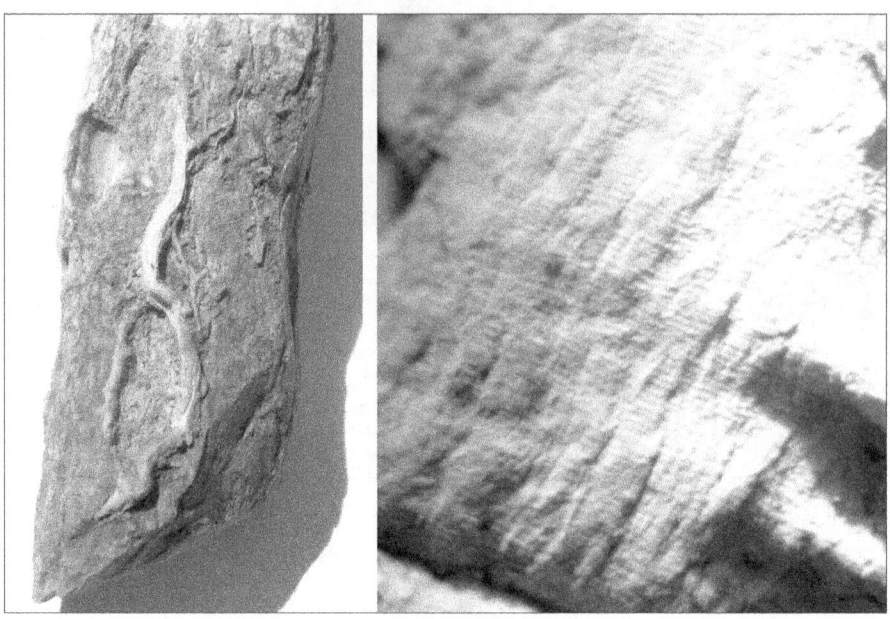

Remarkable surface details are preserved on roots from the Fossil Forest of south west Wiltshire. Rootlets cling to the surface of a large root (left), with a hair-like strand preserved to the right of the main rootlet. Another root surface (right) features transverse parallel lumpy projections and a fine ripple-like pattern cutting transversely across a longitudinal bark-like texture. Specimens: Author's collection. Photos: Author, Steve Clifford.

although cycadophytes also produced 'small scalelike leaves' called cataphylls.[306] In bennettites scaly bisexual cones formed within the armour and probably 'did not extend beyond the level of the trunk surface',[307] whereas in 'true cycads' the cones emerge from the top of the trunk on stalks.

Following the removal of the 0.5 to 0.75 tonne stump of the tree with the aid of a machine, an *in situ* cycadophyte was discovered lying about 1m to the south. The palaeosol sloped down towards the position of the stump at this point, and the fossil was at first not

---

306 Taylor, Taylor and Krings (2009), 703.
307 Taylor, Taylor and Krings (2009), 728.

readily recognisable. Damaged at the edge by the machine, and made up of numerous pieces of varying sizes covered in palaeosol and debris from overlying beds, it was nonetheless excavated in around fifty pieces. During this process distinctive cycadophyte features were recognised, including small areas of diamond-shaped ramenta, and as a result some care and attention was given to cleaning and restoring the specimen as nearly as possible to its original state.

The finished product was not a typical pineapple-shaped Purbeck *Cycadeoidea*, but resembled more a layer cake in which the top layer had slipped sideways. It was also a good-sized specimen, weighing some 40kg (see illustration in preceding chapter). The slightly convex upper surface measured between 0.4m and 0.5m across, with a girth (estimated to include the small missing area) of about 1.45m. In contrast, the specimen from the Isle of Portland named by Seward in 1897 as *Cycadeoidea gigantea* reportedly had a girth of 1.7m and was 0.41m by 0.19m in cross-section.[308] The more elliptical nature of the Dorset fossil suggests that it had fallen over and become laterally compressed before silicification, whereas the Wiltshire specimen had remained standing and become vertically compressed – its maximum height, in fact, reached no more than 0.25m, compared with the 1.18m of Seward's *Cycadeoidea gigantea*. It was also large in terms of diameter when compared with cycadeoids from the Morrison Formation of the western USA – these do not reach 0.5m, although larger trunks 'to 0.51 meter in diameter and 2.4m meters long, occur in Early Cretaceous sediments'.[309]

About four-fifths of the upper surface of the Wiltshire cycadophyte was covered in a scaly mat, formed of flattened ramenta and possibly cataphylls. The remainder of the surface included four small depressed areas of ramenta revealing hollow diamond-shaped

---

308 Cleal, Thomas and Batten (2001), 110.
309 Dayvault and Hatch (2005), 431–2.

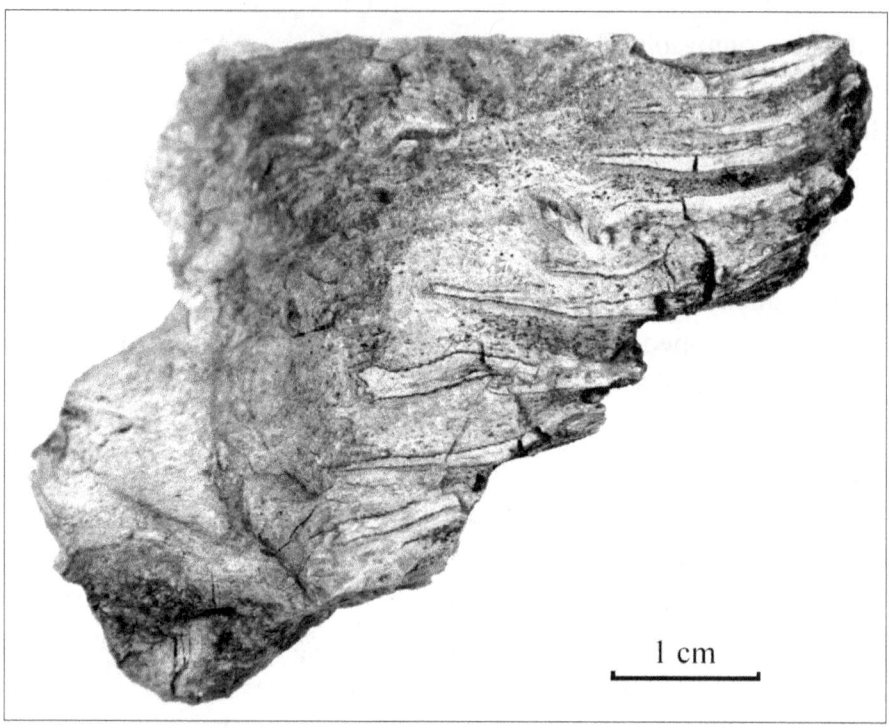

A thin fragment from the largest cycadophyte collected by the author, excavated near the stump of the fossil tree. It comes from halfway up the outer edge of the trunk and exposes internal detail in vertical section. The structures exposed, with slightly saw-toothed edges, are probably ramenta enclosing petioles or leaf stem bases. Specimen: Author's collection. Photo: Steve Clifford.

leaf bases. These could have been areas from which fronds were growing when the plant met its end, or could have been exposed as a result of differential decay and silicification of the trunk.

The layer-cake effect of this cycadophyte appeared to have resulted from two stages of growth. The lower layer, with a maximum height of 0.15m, formed an almost complete donut-shaped ring with a diameter of 0.41m and a central basal projection that presumably was a root base. The upper layer, up to 0.12m in height, had at some stage grown or slipped laterally relative to the lower layer, as a result of which the specimen measured 0.64m along the axis of maximum dislocation. This distortion in growth was in the direction of the downhill slope of the palaeosol, although it could of course have resulted from compression before silicification rather than during growth. Whatever the reason, the fossil has lost the typically elegant sub-spherical to pineapple shape of the best preserved cycadeoid specimens from the Isle of Portland and the Morrison Formation, and is left with a mis-shapen but unusually interesting appearance.

During reassembly of this fossil, the ramenta of the slightly domed top layer of armour were clearly visible along clean breaks, but no underlying structure was visible without thin sectioning or scanning. A small piece from the outer edge of the lower donut-shaped ring revealed detailed internal structure, possibly of the bases of ramenta surrounding petiole bases. The cones that would unequivocally confirm this as a bennettite have not been identified in the armour, but on the basis of overall morphology and other Purbeck bennettite finds it seems highly unlikely it could be other than a bennettite – indeed not only the Dorset Purbeck cycadophytes, but 'most of the Mesozoic cycads [i.e. cycadophyte trunks] from North America and South America available today' belong within this group.[310]

Removal of the tree stump had revealed the largest of the cycadophytes collected by the author, and it was also to reveal the one with the best preservation, illustrated in the Introduction. A single piece of this was found about a metre to the west of the stump, and thorough searching brought to light a further six pieces that led to the reconstruction of a specimen about two-thirds complete. With a diameter of 23cm and a maximum height of 6cm at the apex, it has a helix of diamond-shaped leaf bases radiating out clockwise and anti-clockwise from the centre of the under surface. Between 3 and 6cm out from the centre what appear to be scale-covered buds, possibly

---

310 Daniels and Dayvault (2006), 294.

cone buds, fill some of the ramenta. There is some movement on two of these, like loose teeth in their sockets. Buds are not unknown on fossil cycadophytes, and Seward found on his type specimen of *Cycadeoidea gigantea* from the Isle of Portland a small bud that he thought 'might be an aborted fertile shoot'.[311] The helix disappears into a scaly zone of compressed ramenta, until around 9cm from the apex there lies the inner edge of a donut-like ring extending outwards to the rim. This area is covered in flattened ramenta with fine surface detail preserved. The upper surface is scaly and almost flat, and is marked by a few rounded depressions with diameters of between 1 and 1.5cm that possibly mark the positions of cones.

In this cycadophyte the pith appears to have been completely squeezed out and the armour spread out as one single flattened layer with no vascular tissue preserved. In illustrations of polished sections of compressed specimens from the Morrison Formation[312] a flame effect is sometimes produced as upper and lower layers of armour collapse down onto each other, the 'flames' radiating out from the centre as a double layer of flattened ramenta. The description is particularly appropriate in some American specimens, where minerals in the silica have produced beautiful bright red and yellow to orange colouring. This Wiltshire specimen has either collapsed in a different manner for reasons unknown or was different in its growth pattern.

Another specimen collected from disturbed ground several metres to the south of the stump was reassembled from four pieces and was over 75 per cent complete. It formed an approximate circle with a diameter averaging 26cm, had a maximum height of 6cm, and had none of the distinctive diamond-shaped leaf bases on either the slightly concave scaly upper surface or the under surface. Once again it revealed evidence of armour but not of pith along radial breaks,

---

311 Cleal, Thomas and Batten (2001), 112.
312 Dayvault and Hatch (2005), Figs 9 and 10.

A highly compressed 26cm diameter cycadophyte surface collected from the Purbeck Fossil Forest of south-west Wiltshire. The missing segment exposes a 'flame' effect caused by compression of the ramenta, which have collapsed to leave no obvious evidence of conductive tissue or pith. Specimen: Author's collection. Photo: Steve Clifford.

but in this case a flame outline was readily discernible. However, the effect appeared to have been caused by a concertina-style collapse of a single layer of armour, with no preserved evidence of armour at the top and bottom of the trunk squeezing out intervening pith and vascular tissue. A stalk base with a diameter of about 2mm was visible in one of two incomplete concentric rings of small round depressions on the underside, along with two small indeterminate bud-like structures.

A further specimen, about half complete and collected from surface fragments, had a diameter of 35cm and a maximum height of 6cm, producing a width to height ratio of almost six to one. The flame structure was once again present, but less pronounced and restricted to the rim area of the specimen. It could be the case that armour along the top of the cycadophyte was relatively resistant to vertical compression, but that the armour down the sides was prone to complete collapse from the same level of compression, which only leaves the question as to why similarly compressed specimens are not to the author's knowledge found elsewhere. The top was covered with flattened ramenta and there were scattered round depressions on the underside with an average diameter of around 2cm. From radial breaks in the fossil these could be seen to lie at the base of structures within the trunk, possibly cones, which would confirm that the specimen was a bennettite.

One other specimen was represented by a single fragment, making six in all including the one attached to the tree. Although distribution of these could not be mapped precisely because four were found in disturbed ground, they do provide evidence of the abundance such plants, at least on a localised level. The one large and five smaller specimens came from a total area measuring little more than 50 to 60sq.m, and did not necessarily represent the original total growing there.

The discovery of the tree and of the cycadophytes had resulted from initial surface collecting on partially disturbed ground, and the site clearly deserved a more systematic excavation of a trial area. The line of the palaeosol was readily discernible some 15m to the west of the tree stump position, and a cycadophyte fragment and a couple of small conifer fragments had been found on the surface in this area. With a spade to remove bulky overburden, and a trowel for detailed work exposing the top of the bed, work began in the summer of 2010 on the exposure of a 4m by 10m area. A small band of sticky brown clay overlain by a thin layer of crumbly white marl marked the top of the palaeosol, indicating the point at which spadework ended and trowel work began. The palaeosol turned out to have a very uneven surface, to all appearances retaining its original topography rather than having been distorted by later geological movements. A contour map was produced using a laser level and aerosol paint marker spray, and fossil finds were marked onto the map.

Although lengthy periods of excavation resulted in no finds beyond the occasional black carbonaceous streak, when discoveries

A group of four cycadophytes in original growth positions revealed on a sloping surface of exposed palaeosol of the Wiltshire Fossil Forest. Preservation indicates that stronger growth occurred in the direction away from the bank. The specimens were all fragmented and their positions carefully recorded before collection. Specimens: Author's collection. Photo: Author.

did take place they more than made up for the unproductive hours. Silicified specimens represented the standard flora of conifer wood and cycadophytes, but now provided new details of growth and density. Although five complete cycadophytes were found within the collecting area, four of these formed a single group growing on a small bank within an area of under 1sq.m. Two of them showed marked asymmetrical growth in the direction of the downward slope of the ground, and the other two showed a marginal growth distortion in this direction. This reinforced the possibility that irregular growth on the large 'layer-cake' cycadophyte near the tree was a consequence of the uneven ground.

The four cycadophytes in the group ranged from 16 to 26cm in diameter and were all represented by a single layer of armour with diamond-shaped leaf bases only visible in the central area of the under surface. The outer sections of the under surfaces were covered with flattened ramenta. On the upper surfaces the leaf base pattern was visible only on the smallest 16cm diameter specimen, and the other three were covered in a mat of flattened ramenta. The plants appeared to have grown mushroom-style from the ground, and unless they were a parasitic form of cycadophyte the absence of vascular tissue and pith must have been due to incomplete silification. The contrast in shape to Dorset *Cycadeoidea* specimens could have resulted from a different inter-relationship between the processes of decay, compression and silicification, or could have resulted in part at least from these being a different type of cycadophyte. None of these specimens yield any obvious external evidence of cones.

The fifth and largest cycadophyte, some 1.5m from the group, was 35cm in diameter with a maximum height of 7cm. The centre was poorly preserved with no visible leaf bases, the underside had a central projection that probably marked a root base, and flattened ramenta covered the remaining visible surface areas. All five cycadophytes lay within a small part of the 40sq.m excavation, and although it is possible that other examples had decayed rather than become silicified, significant carbonaceous traces such as would most likely be left by these woody structures were not observed.

About 1.5m from the cycadophyte group, on flat ground above the bank on which they had grown, were excavated the remains of a small conifer, probably some eight or nine years old judging by growth rings. Sections of irregularly silicified wood lay on the surface of the palaeosol, including a piece of what was probably a short rounded bole with a diameter of 8cm. Recovered remains indicated three branches emerging from the bole a short distance above ground level. This could have represented the natural growth pattern or could have resulted from pollarding by herbivores – Jurassic conifers were far more likely to be stripped of leaves and branches than today's. Several roots were found *in situ* beneath the surface of the palaeosol, and some bark was preserved on the roots, bole and branches. In the case of two roots the bark preservation was exceptional and they could have acted as buttresses above ground level to support the bole. The roots also gave clear indications of the way in which the conifer anchored itself to the palaeosol. In one case, three minor roots spiralled out from the main root and descended claw-like into the ground. Ridges on the undersides of minor roots indicated that selective contraction had occurred to increase grip. In the other case, the main root descended into the soil to a depth of 11cm before sending two lesser roots downwards and continuing in a horizontal direction. A third root lacked bark and showed a simpler structure, meandering horizontally for some 48cm and tapering from a diameter of 2.5cm to 1cm. Two rootlets of between 3mm and 5mm in diameter were also preserved, and an examination of internal detail at breaks between sections showed clear cell rows and growth rings.

Beneath the conifer root system, a straight root with a distinct swelling or node ran at a depth of between 15 and 22cm beneath the surface of the palaeosol, possibly forming part of a runner or rhizome from a different plant. A 21.5cm section was reconstructed from a number of broken sections with an average diameter of 1.5cm, but at either end the pieces were too poorly preserved for successful reconstruction. Some bark was preserved, and on one small section fell away to reveal minute rootlets attached to the inside of the bark and to the wood surface, in one case penetrating down into the wood.

A 40sq.m excavation had yielded the remains of five cycadophytes and one small conifer within no more than a fifth of that area, but the remainder of the dig proved far less productive. Sections of a conifer root were found along the bottom of the bank on which the cycadophytes grew, and four fragmentary conifer root sections were found at the opposite end of the excavation. The roots, lacking the exceptional preservation of the others possibly because of their pre-fossilisation condition, showed evidence of attached fine

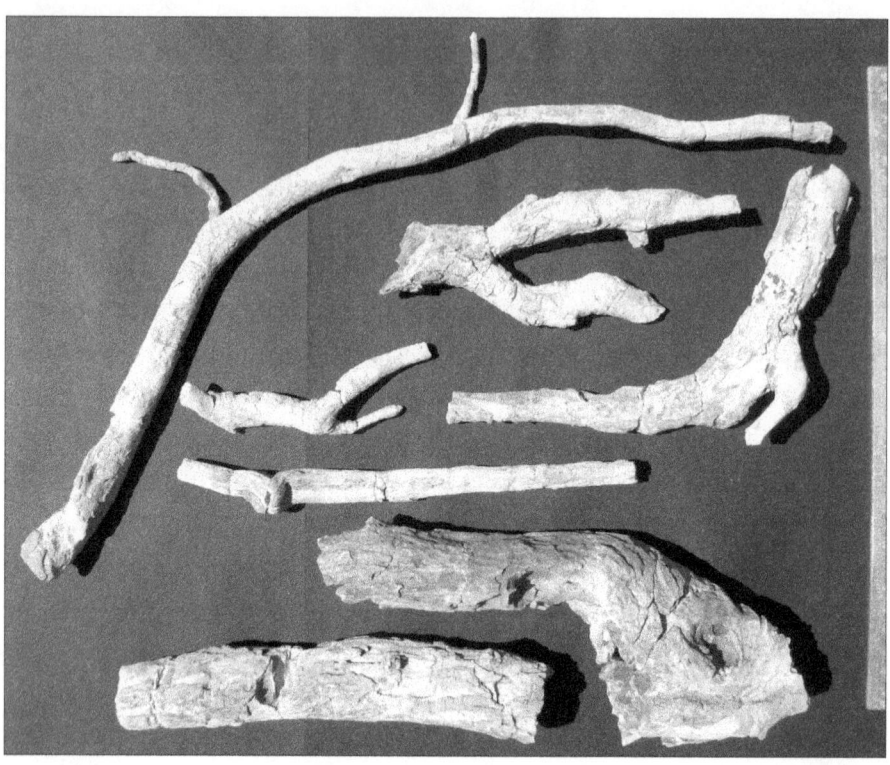

Conifer remains from the Wiltshire Fossil Forest, most if not all from a single tree about eight years old. Elements of the root system (top and centre) include buttress-like specimens that were probably exposed above soil level. The bottom two specimens are elements of the intermittently preserved branching system, with areas of exceptional bark preservation. Specimens: Author's collection. Photo: Author.

rootlets in places – rootlets that were noticeable by their absence on much of the healthy looking root system of the young conifer.

Two bags of palaeosol were removed from around the base of the small conifer and put through an 0.5mm sieve. Fossils of Purbeck insects and vertebrates nearly all come from the middle and upper parts of the Purbeck Limestone Group, and the results of the sieving were something of a surprise. Perhaps of most note was the silica-infilled abdomen of what appears to be an insect or the respective part of a pupa. It seems likely that this was a shed exoskeleton or pupa case, possibly of a wasp or beetle,[313] but whatever it was the three-dimensional petrification of the remains of a bug is remarkable. Other finds included small snails, a small piece of possible silicified algae like a miniature stromatolite, two teeth, three pieces of reptile eggshell, and some indeterminate material.

The small teeth, less than 2mm in length, are typical of those from small crocodiles, probably juveniles of the 'dwarf' genus *Theriosuchus*. With similar crocodile teeth found in the plant and reptile bed and the Purbeck Limestone Group at Durlston Bay, they are another missing link in a complex story. The eggshell fragments require further examination. Surface details are not identical to those of the eggshell found in the plant and reptile bed, but subject to more specific identification it is tempting to imagine diminutive crocodiles emerging from their eggs beneath the foliage of extinct cycadophytes and conifers.

All in all, the excavation had been remarkably successful. Here was a detailed snapshot of an Upper Jurassic terrestrial environment, with the soil topography mapped, with *in situ* cycadophytes and a small conifer, with evidence of snails and bugs and with evidence of one reptile species. Whether or not any of these specimens had been transported into the palaeosol is an unanswerable question, but their delicacy, especially in the case of the insect, indicates that they could well be autochthonous. There is evidence that eggshell 'can be transported a considerable distance with minimal abrasion',[314] but there is a good chance that eggshell found within a vegetated area on a low bank could be there because eggs were laid in the vicinity.

Before the above systematic excavation had taken place, the tops of some limestone stromatolites had been observed some 65m to the east of the tree stump, and a small area had been uncovered to

---

313   Munt, M. (2011) Pers. comm.
314   Ensom (1997), 83.

Fossils sieved from palaeosol collected around the young Fossil Forest conifer illustrated previously. Provisional identifications, clockwise from top left, are as follows: Insect abdomen; coprolite; gastropod; reptile eggshell; and crocodile teeth cf. Theriosuchus. A 50kg sample was sieved to 0.5mm. Specimens: Author's collection. Photo: Author.

reveal a number of cauliflower-shaped structures sitting on an apparently eroded and fractured limestone platform. The exact relationship between this horizon and the Fossil Forest horizon had not been determined, and a small exploratory area was excavated. This revealed that, as elsewhere, the palaeosol was variable in thickness and the surface uneven. It was found to abut the limestone platform at one spot, and to rise over the edge of it at others. It was discontinuous, and from the available evidence it appeared that the limestone platform with stromatolites would have been partially exposed adjacent to areas of soil formation. The platform also helped explain the irregular surface of the palaeosol as a possible feature of karst topography,[315] or landscaping resulting from the dissolution of limestone – in more advanced stages this leads to cave formation and underground river courses.

The excavation led to some fortuitous finds. Evidence indicated that fossils found so far in the Wiltshire Fossil Forest had been collected either from positions of growth, from positions onto which branches and trunks had directly fallen, or from recently disturbed surface material. Along the edge of the limestone platform and up to a metre or more from it there turned out to be a large amount of silicified conifer wood and cycadophyte remains that lay randomly scattered around. Some specimens showed considerable signs of wear,

---

315 Munt, M. (2010) Pers. comm.

and sections of wood, ranging from twigs to medium-sized branches, were seldom more than 30cm in length. There seemed little doubt that this was a layer of forest debris incorporating material in varying states of decay that had been deposited against the limestone edge.

One cycadophyte specimen, some 17cm by 20cm, had been laterally compressed to a thickness of 5cm – this dimension included two layers of armour, vascular tissue and any remnants of the pith, and reinforced evidence from *in situ* specimens that the cycadophytes in this area had all been highly compressed. By way of contrast, the ratio of minimum to maximum diameter on some conifer wood debris was no more than 1:1.5. The diamond-shaped ramenta on the laterally compressed cycadophyte specimen were readily visible and far larger than those on any other specimens collected, with average widths of 3cm and heights of 2cm. A few other cycadophyte fragments came from vertically compressed specimens within the 20 to 30cm diameter range, and two were from small cycadophytes probably less than 12cm in diameter with delicately layered paper-thin ramenta.

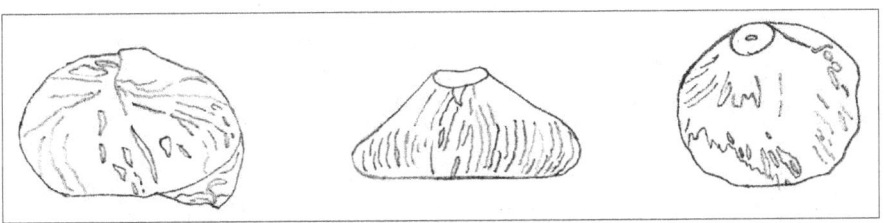

Small lignotubers and/or Steinerocaulis-type short shoots from the Tithonian of south west Wiltshire: 12mm diameter specimen from Wiltshire's Fossil Forest (left, author's collection) and plant and reptile bed specimens (centre, dia. 8mm, NHM V.65150; right, dia. 14mm, NHM V.65123). Drawings: Author, plant and reptile bed specimens from Needham (2007), Pls 2 and 3.

Most of the conifer wood specimens revealed little detail of scientific significance, although several of them possessed a certain aesthetic appeal. A few rare examples demonstrated branching at acute angles and multiple branching from small nodes – in one case four truncated twigs were located round one node. Some possible bark was found on a few specimens, although no evidence of scaly foliage on any of the twigs.

The real treasures of this silicified detritus were to be found in three small specimens. The first was a 12mm diameter rounded fossil with a height of 6mm, similar to a lignotuber or to a *Steinerocaulis*-like short shoot from the Morrison Formation or the plant and reptile bed. Although much smaller, in terms of morphology and relative dimensions it bears a remarkable similarity to a 'broken seedling' (NHM V.30948) from the Cerro Cuadrado Petrified Forest of Patagonia,[316] a fossil form now re-interpreted as possible lignotubers.[317] The truncated stalk from which it had grown was discernible on the centre of the underside, and along one side a piece of the twig or small branch from which it had grown was compressed against it. By any standards it was a remarkable piece of detail from the Purbeck Fossil Forest, a missing link from rare forest debris preserved close to *in situ* vegetation.

The other two small specimens were conifer cones, a mere 10mm and 14mm across but with some exceptionally well-preserved detail. They were nearly overlooked among the assorted debris of wood fragments, chert pebbles and assorted semi-angular limestone pebbles found in this part of the palaeosol, and it was only when cleaned and put under a magnifying lamp that their true significance could be seen. These were to the author's knowledge the first silicified cones from the Purbeck Fossil Forest and the first female cones, showing prominent fertile seed-bearing scales and the subtending scale-like bracts. At the base of the smaller cone sterile foliage compressed against the base had a finely striated surface, a feature noted in foliage at the base of Patagonian *Paracauraria* cones, and also common to these cones was longitudinal ridging along the protruding surface of the fertile

---

316 Calder (1953), Pl. 2, Fig. 17.
317 Stockey (2002), 171.

scales.[318] Preservation of such detail suggests that these were freshly fallen cones when they were deposited among the other debris.

	Wilts 'plant and reptile bed'	Wilts Fossil Forest	Dorset Forest/ Charophyte Cht	Dorset Purbeck Berriasian	Morrison Formation, USA	Cerro Cuadrado Argentina
In situ conifer fossils		•	•	•		•
B. provoensis-type shoots	•	•			•	
B. joannei-type shoots	•				•	
S. radiatus small peltate shoots	•				•	
Lignotubers and/or similar *Steinerocaulis*	•	•			•	•
Carpolithes westi/ *Araucarites*? scales	•		•		•	
Araucarites/Araucaria cones			•			•
C. acinus, C. gibbus, C. rubeola, C. glans	•		•			
Cycadophytes Undifferentiated		•	•		•	
Turtle: Solemydidae Undifferentiated	•			•		
Crocodile: *Goniopholis* sp.	•			•	•	
Crocodile: Small/ dwarf forms	•	•		•	•	
Sauropod teeth: *Camarasaurus*/similar	•				•	
Mammal assemblages	•			•	•	

Table 6: Distribution table of selected Wiltshire Tithonian fossils.

318 Calder (1953), 120.

Despite these common features with *Pararaucaria* cones, there are significant differences and it is possible they are cheirolepidiaceous cones, from the same family as the tree *Protocupressinoxylon purbeckensis*. If so, they are rare finds – although a widespread family in Mesozoic times, 'there is not a great deal of information about the seed cones of the Cheirolepidiaceae'.[319] With modern CT scanners available the cones can be examined for internal structure without being sectioned, and hopefully enough structure will be preserved for them to be identified. Although no direct attachment has been found, there is a strong association between the cones and the tree, but if more than one type of tree were growing in close proximity there are obvious dangers in jumping to conclusions.

From a branching tree trunk over 11.5m in length to a clump of cycadophytes growing on a bank beneath a small conifer, and from an undulating soil surface to an eroded stromatolite-covered limestone platform with plant debris deposited along its rim, the Wiltshire Fossil Forest has more than revealed its value. It is a big piece of the jigsaw, adding in new ways to the overall picture of the Tithonian forests of southern England and linking them along with the plant and reptile bed material to the forests of the western USA and Patagonia. The true scientific work lies ahead,

One of two rare small silicified cones found in the palaeosol of Wiltshire's Fossil Forest amidst debris adjacent to a limestone platform. These are possible examples of little-known female Cheirolepidiaceae cones, and CT scanning at the NHM will hopefully reveal any preserved anatomical detail. Specimen: NHM. Photo: NHM.

319 Taylor, Taylor and Krings (2009), 835.

and in addition to the fossils described in these pages, samples have been taken by the NHM from significant horizons and full geological sections recorded.

As one who regrets the increasing demise of traditional museum displays, and who remembers with some nostalgia the fine displays of Devonian fish and of crocodile skulls collected by the Marchioness of Hastings in the first half of the nineteenth century, now consigned to the backrooms of the NHM, the author retains the hope that the fossil tree described in this chapter will be put on public display and not consigned to indefinite storage. Of the other material from south-west Wiltshire, some has been donated to the museum and further specimens will be offered to them, especially where they are of importance in furthering scientific knowledge. The author has always attempted to collect, record and notify specimens to a high standard within the practical constraints of time and of site conditions, and reckless collecting would have destroyed the scientific value of the material, as has happened at the Cerro Cuadrado where 'large scale removal of spectacular fossil cones ... has left us without an appreciation for the placement of these specimens within the sediment'.[320] Fossils have been treated as part of the scientific heritage to which they belong.

---

320 Stockey (2002), 171.

Postscript

A WORLD OF VAST geological time scales had taken hold of the public imagination many years before Darwin's *Origin of the Species* was published in 1859. The process was helped along not only by pioneering geologists but also by the poet laureate Alfred Tennyson. Two famous and often quoted verses from his 1851 poem *In Memoriam*, the most widely read poem of its day, put it eloquently:

> There rolls the deep where grew the tree.
> O earth, what changes hast thou seen!
> There where the long street roars, hath been
> The stillness of the central sea.
>
> The hills are shadows, and they flow
> From form to form, and nothing stands;
> They melt like mist, the solid lands,
> Like clouds they shape themselves and go.

Looking south from Tisbury High Street, the hills that rise up on the far side of the River Nadder tell their story. Travel back 65 million years to the extinction of the dinosaurs and they weren't there, although the rocks of which they are made were there, waiting to be thrust upwards by the folding of the Earth's crust. Throughout much of the Upper Cretaceous, the final epoch of the dinosaurs, a good part of England and a great swathe of Europe lay beneath a calm warm sea. The climate was extremely hot, of the type threatened for humans by the severest of global warming predictions. White chalk deposits, formed on this sea bed over tens of millions of years from the skeletons of microscopic plankton, fringe the northern and southern rims of the Vale of Wardour and make up Salisbury Plain and Cranborne Chase. They have been eroded away from the centre of the Vale of Wardour, broken up by frosts and dissolved by millions of years of rain and river action.

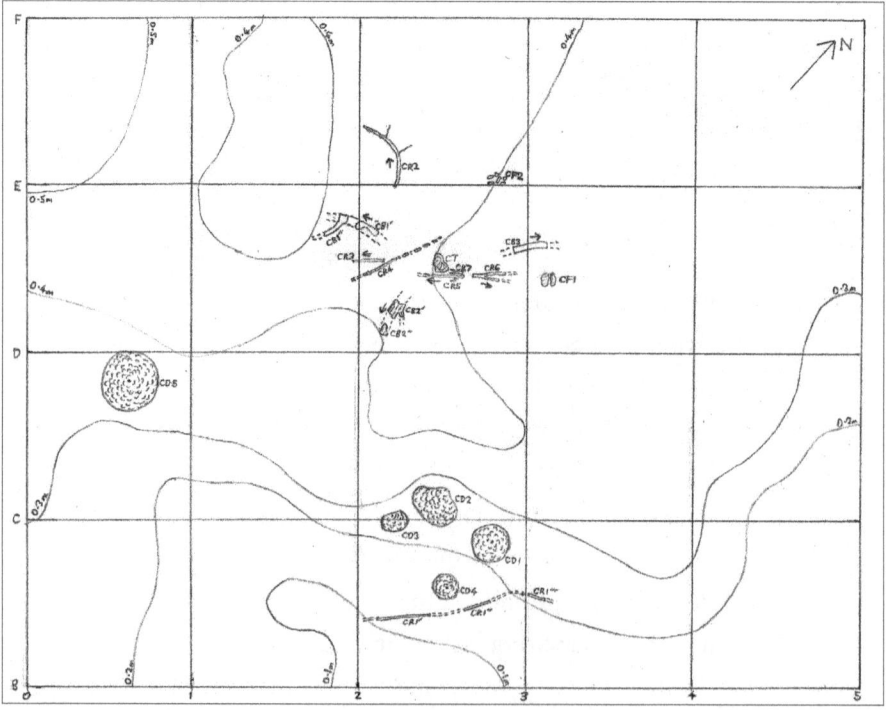

Grid map of a 20sq.m area of excavated palaeosol from Wiltshire's Fossil Forest, contoured at 10cm intervals. CD1 to CD5 are cycadophytes, CT is a probable conifer trunk base, CB1 to CB3 are sections of conifer branches, CR1 to CR7 are conifer roots, and CF1 and CF2 are fragmentary conifer material. Dotted lines are conjectural and arrows indicate directions of growth. Map: Author.

✻ Postscript ✻

Below the chalk lie Lower Cretaceous deposits of green sand and sandstone. This is the Upper Greensand, whose building stones have been widely used in some parts of the vale and in nearby Shaftesbury. Land was near enough for current action to transport river-born sand grains out onto the sea bed. Hard weather-resistant beds of chert, believed to have been formed from the siliceous spicules of sponge skeletons, cap the Upper Greensand. As a result it now forms ridges of high hills within the chalk escarpments of the vale, heavily wooded on sandy well-drained soils and the site of forts and settlements dating back to pre-Roman times.

By now the Age of Flowers has been left behind, as the journey back into the Lower Cretaceous continues through the marine Gault Clay to the Lower Greensand, from which on the Isle of Wight the earliest British angiosperms have been collected – the remains of plants carried out to sea before sinking waterlogged onto the sea floor. Below the Lower Greensand lie several Wealden-type formations of the Wessex and Weald basins, extending down to the top of the Purbeck Limestone Group and and covering a period of some 15 million years. In south-west Wiltshire the Wessex Formation is only represented by thin clays towards the eastern end of the Vale of Wardour around Dinton.

If poorly represented in Wiltshire, these Wealden-type deposits are well developed elsewhere and tell the story of low-lying deltaic, lagoonal, lacustrine and wetland conditions across south-central England before the marine inundation of the Lower Greensand. The Isle of Wight is famous for its dinosaur fossils from the Wessex Formation, found in association with plant remains. Plants are also found on the mainland, and a notable site on the Sussex coast reveals the diversity of the flora: quillworts; horsetails of the genus *Equisetum*; one species each of the pteridospermophyte orders Caytoniales and Corystospermales; representatives of ten families of ferns; one genus of Order Czekanowskiales; several leaf genera of cycadophytes of Orders

Cycadales and Bennettitales; representatives of five determinate conifer families, the Cheirolepidiaceae, Araucariaceae, Taxodiaceae, Pinaceae and Taxaceae; and possible leaves of Order Ginkgoales.[321] A possible but unconfirmed angiosperm has also been described from the Wealden Clay.[322]

By now England lies some 15 degrees further south than at the start of the time travel, and the broad Atlantic Ocean is a narrow sea. The Wealden-type deposits give way to the Purbeck Limestone Group, with lagoonal and terrestrial deposits in Dorset and Wiltshire. Some environmental conditions persist between the Wealden and Purbeck times, but others have changed as evidenced by the different geological deposits, fossil assemblages and taphonomic processes. Somewhere in the lower part of the group, about 145.5 million years back in time, the transition into the Upper Jurassic is made – a human artifice of little concern to the dinosaurs.

Not far into the Upper Jurassic of Wiltshire, maybe around 146mya in the latter part of the Tithonian Age, there is a remarkable opportunity to stop and take in some details of a terrestrial scene – to sit on a shallow bank by a branching coniferous shrub, with a small group of cycadophytes growing at one's feet, the palm-like fronds growing out from the centres of the armoured trunks. Small land snails and insects are to be seen, and small crocodiles run off in alarm. Unseen animals, from the small to the enormous, could appear at any moment. Scattered trees dot the surrounding area, conifers up to 15m or more tall that branch and rebranch. Near to their bases and scattered around are more cycadophytes, the predominant feature of the undergrowth. Beyond the nearest tree the soil gives way to an eroded platform of bare limestone, covered in cauliflower-like mounds of stromatolites that formed when the area was under

---

321 Thomas and Batten (2001b), 145–6.
322 Thomas and Batten (2001b), 141.

* Postscript *

An artist's reconstruction of the south-west Wiltshire environment during late Tithonian times, when the Fossil Forest palaeosol was formed. Cycadophytes on hummocky soil and conifer trees give way to a stromatolite covered limestone platform and, in the distance, sauropods crossing broad mudflats and the shallow waters of a lagoon. Drawing: Isabelle Needham.

water. An assortment of forest floor debris lies along the edge of the limestone.

Carry on back further in time and the land sinks beneath the waves once more, firstly of a lagoon and then of the warm shallow Portland sea. For a geologically brief period of time lagoonal and terrestrial conditions return. On the floor of the lagoon plant and animal debris swept in from adjacent land lies buried in a bed of silt. The silt fills a depression or degraded channel eroded into a former sea bed that had become dry land. Fragments of the big picture can be reassembled from these scattered remains of past life, assisted by comparisons with fossils from Dorset and the Americas: large sauropods crossing coastal plains, possibly in herds; fierce serrated-toothed hunters; small and large crocodiles hunting in and around shallow lagoons; pterosaurs soaring overhead; lizards and small mammals rustling in the undergrowth; a flora dominated by branching conifer trees and cycadophytes but also, according to evidence from various associated assemblages, including lycopods, horsetails, ferns, pteridospermophytes, other conifers and ginkgophytes, some possibly similar to plants found in the Middle Jurassic of Yorkshire and in the Wealden-type floras. A single bed of well-preserved leaf compressions within the Purbeck Limestone Group could answer so many questions, and yet such are the vagaries of the fossil record that no such bed has been discovered – and even though miospore assemblages have been examined, the sorting and differential decay of such material means that even 'a high concentration of dispersed pollen does not necessarily reflect local macrovegetation'.[323]

Travelling home through time from this strange and alien Jurassic landscape, the traveller must to be careful to stop at exactly the right moment in order to return to Tisbury. The village and its life come and go in the flickering of a single frame of geological time,

---

323 Francis (1983), 292.

✳ Postscript ✳

before the hills around it enter another epoch and flow onwards to assume new forms, and before the creatures that inhabit it go the way of the pterosaurs, the sauropods, the theropods, and so many others of the creatures both strange and familiar that lived in its environs around 146 million years ago.

REFERENCES

Agashe, S.N. (1997) *Paleobotany: Plants of the past, their evolution, paleoenvironment and application in exploration of fossil fuels*, Science Publishers, Inc.

Anon. (1983) Report on new excavations at Chicksgrove, Wiltshire. Unpublished report for the Nature Conservancy Council.[324]

Astin, T.R. (1987) Petrology (including fluorescence microscopy) of cherts from the Portlandian of Wiltshire, UK – evidence of an episode of meteoric water circulation. In Marshall, J.D. (ed.) *Diagenesis of Sedimentary Sequences*, Geological Society of London Special Publications, **36**, 73–85.

Barker, D., Brown, C.E., Bugg, S.C. and Costin, J. (1975) Ostracods, land plants, and Charales from the basal Purbeck Beds of Portesham Quarry, Dorset. *Palaeontology*, **18**, 419–36.

Barrett, P. (2010) Digging for Dinosaurs. *Evolve*, **4**, 58–61.

Bartholomew, M. and Morris, P. (1991) Science in the Scottish Enlightenment. In Goodman, D. and Russell, C.A. (eds) *The Rise of Scientific Europe 1500–1800*, Hodder and Stoughton.

Baskin, C.C. and Baskin, J.M. (1998) *Seeds: Ecology, Biogeography, and Evolution of Dormancy and Germination*, Academic Press.

Bass, C. (1964) Significant new plant locality in Utah. *Contributions to Geology*, **3**, (2): 94–5.

Batten, D.J. (2002) Palaeoenvironmental setting of the Purbeck Limestone Group of Dorset, southern England. In Milner, A.R. and Batten, D.J. (eds) *Life and Environments in Purbeck Times*, Special Papers in Palaeontology No. 68, The Palaeontological Association, London.

---

[324] Entry as in Benton, Cook and Hooker (2005), 164.

Bennett, G.E. (2008) *Faunal Diversity in Paleoecosystems: A Model for Using the Species-Area Relationship to Analyze Paleoenvironments*, Thesis submitted to George Mason University, Fairfax, VA, USA.

Benton, M.J., Cook, E. And Hooker, J.J. (2005) *Mesozoic and Tertiary Fossil Mammals and Birds of Great Britain*, Geological Conservation Review Series, No. 32, Joint Nature Conservation Committee, Peterborough.

Benton, M.J., Hooker, J.J. and Cook, E. (2005) British Mesozoic fossil mammal GCR sites. In Benton, M.J., Cook, E. And Hooker, J.J. (2005) *Mesozoic and Tertiary Fossil Mammals and Birds of Great Britain*, Geological Conservation Review Series, No. 32, Joint Nature Conservation Committee, Peterborough

British Geological Survey (1999) *The Wincanton district – a concise account of the geology*, London: The Stationery Office.

Brown, C.E. and Bugg, S.C. (1975) The Land Plants. In Barker, D., Brown, C.E., Bugg, S.C. and Costin, J. Ostracods, land plants, and Charales from the basal Purbeck Beds of Portesham Quarry, Dorset. *Palaeontology*, **18**, 419–36.

Brown, L. (ed.) (1993) *The New Shorter Oxford English Dictionary*, Oxford: Clarendon Press

Brunsden (ed.) (2003) *The Official Guide to the Jurassic Coast*, Wareham: Coastal Publishing.

Buffetaut, E. (1990) A sauropod dinosaur in the Portlandian of Haute-Marne (Eastern France). *Geobios*, **23**, 6, 755–60.

Cadbury, D. (2001) *The Dinosaur Hunters* London: Fourth Estate.

Calder, M.G. (1953) A Coniferous Petrified Forest in Patagonia. *Bulletin of the British Museum (Natural History) Geological series*, **2**, 99–138, pls 1–7.

Chandler, M.E.J. (1966) Fruiting Organs from the Morrison Formation of Utah, U.S.A.. *Bulletin of the British Museum (Natural History) Geological series*, **12**, 137–71, pls 1–12.

Cleal, C.J. and Thomas, B.A. (2001) Introduction to the Mesozoic and Tertiary Palaeobotany of Great Britain. In Cleal, C.J., Thomas, B.A., Batten, D.J. and Collinson, M.E. *Mesozoic and Tertiary Palaeobotany of Great Britain*, Geological Conservation Review Series, No. 22, Joint Nature Conservation Committee, Peterborough.

Cleal, C.J., Thomas, B.A. and Batten, D.J. (2001) The Jurassic palaeobotany of southern England. In Cleal, C.J., Thomas, B.A., Batten, D.J.

and Collinson, M.E. *Mesozoic and Tertiary Palaeobotany of Great Britain*, Geological Conservation Review Series, No. 22, Joint Nature Conservation Committee, Peterborough.

Cleal, C.J., Thomas, B.A., Batten, D.J. and Collinson, M.E. (2001) *Mesozoic and Tertiary Palaeobotany of Great Britain*, Geological Conservation Review Series, No. 22, Joint Nature Conservation Committee, Peterborough.

Collinson, M.E. and Cleal, C.J. (2001a) Early and early middle Eocene (Ypresian-Lutetian) Palaeobotany of Great Britain. In Cleal, C.J., Thomas, B.A., Batten, D.J. and Collinson, M.E. *Mesozoic and Tertiary Palaeobotany of Great Britain*, Geological Conservation Review Series, No. 22, Joint Nature Conservation Committee, Peterborough.

Collinson, M.E. and Cleal, C.J. (2001b) Late middle Eocene-early Oligocene (Bartonian-Rupelian) and Miocene palaeobotany of Great Britain. In Cleal, C. J., Thomas, B.A., Batten, D.J. and Collinson, M.E. *Mesozoic and Tertiary Palaeobotany of Great Britain*, Geological Conservation Review Series, No. 22, Joint Nature Conservation Committee, Peterborough.

Cope, J.C.W., Ingham, J.K. and Rawson, P.F. (eds) (1992) *Atlas of Palaeogeography and Lithofacies*, Geological Society Memoir No. 13, The Geological Society, London.

Cope, J.C.W., Rawson, P.F. and Wimbledon, W.A. (1992) J11a-d: Portlandian. In Cope, J. C. W., Ingham, J. K. and Rawson, P. F. (eds.) (1992) *Atlas of Palaeogeography and Lithofacies*, Geological Society Memoir No 13, The Geological Society, London.

Daniels, F.J. And Dayvault, R.D. (2006) *Ancient Forests: A Closer Look at Fossil Wood*, Western Colorado Publishing Company, Grand Junction.

Dantas, P., Sanz, J.L., Da Silva, C.M., Ortega, F., Dos Santos, V.F. and Cachão, M. (1998 trans. Harris, J.D. 2002) Lourinhasaurus n. gen. A new sauropod dinosaur from the Upper Jurassic (Upper Kimmeridgian-Lower Tithonian) of Portugal. *V Congreso Nacional de Geologia, Lisboa*, 91–4.

Dayvault, R.D. and Hatch, H.S. (2003) Short Shoots from the Late Jurassic Morrison Formation of Southeastern Utah. *Rocks and Minerals*, **78**, 232–51.

Dayvault, R.D. and Hatch, H.S. (2005) Cycads from the Upper Jurassic and Lower Cretaceous Rocks of Southeastern Utah. *Rocks and Minerals*,

80, 412-32.

Delair, J.B. and Wimbledon, W.A. (1995) Reptilia from the Portland Stone (Upper Jurassic) of England: A Preliminary Survey of the Material and the Literature. In *Vertebrate Fossils and the Evolution of Scientific Concepts*, Sarjeant, W.A.S. (ed.), Gordon and Breach Publishers.

Dernbach, U. and Tidwell, W.D. (eds) (2002) *Secrets of Petrified Plants: Fascination from Millions of Years*, D'ORO Publishers, Heppenheim.

Ensom, P.C. (1988) Excavations at Sunnydown Farm, Langton Matravers, Dorset: amphibians discovered in the Purbeck Limestone Formation. *Proceedings of the Dorset Natural History and Archaeological Society*, **108**, 205-6.

Ensom, P.C. (1997) Reptile eggshell from the Purbeck Limestone Group of Dorset, southern England. *Proceedings of the Dorset Natural History and Archaeological Society*, **118**, 79-83.

Ensom, P.C. (2002) The Purbeck Limestone Group of Dorset, southern England: A guide to lithostratigraphic terms. In Milner, A.R. and Batten, D.J. (eds) *Life and Environments in Purbeck Times*, Special Papers in Palaeontology No. 68, The Palaeontological Association, London.

Escaso, F., Francisco, O., Dantas, P., Malafaia, E., Pimentel, N.L., Pereda-Suberbiola, X., Sanz, J.L., Kullberg, J.C., Kullberg, M.C., Barriga, F. (2007) New evidence of shared dinosaur across Upper Jurassic Proto-North Atlantic: Stegosaurus from Portugal. *Naturwissenschaften*, **94**, 367-74.

Evans, S.E. and McGowan, G.J. (2002) Lissamphibian remains from the Purbeck Limestone Group, southern England. In Milner, A.R. and Batten, D.J. (eds) (2002) *Life and Environments in Purbeck Times*, Special Papers in Palaeontology No. 68, The Palaeontological Association, London.

Evans, S.E. And Searle, B. (2002) Lepidosaurian reptiles from the Purbeck Limestone Group of Dorset southern England. In Milner, A.R. and Batten, D.J. (eds) *Life and Environments in Purbeck Times*, Special Papers in Palaeontology No. 68, The Palaeontological Association, London.

Fahn, A. (1990) *Plant Anatomy* (4th ed.), Pergamon Press.

Falcon-Lang, H.J. (2004) A new anatomically preserved ginkgoalean genus from the Upper Cretaceous (Cenomanian) of the Czech Republic.

References

Palaeontology, **47** (2), 349–66.

Francis, J.E. (1983) The dominant conifer of the Jurassic Purbeck Formation. *Palaeontology*, **26**, 277–94.

Francis, J.E. (1984) The seasonal environment of the Purbeck (Upper Jurassic) fossil forests. *Palaeogeography, Palaeoclimatology, Palaeoecology*, **48**, 285–307.

Geddes, I. (2000) *Hidden Depths: Wiltshire's Geology and Landscapes*, Ex Libris Press, Bradford-on-Avon.

Gnaedinger, S. (2007) Podocarpaceae woods (Coniferales) from middle Jurassic La Matilde formation, Santa Cruz province, Argentina. *Review of Palaeobotany and Palynology*, **147**, (1–4), 77–93.

Gradstein, F.M., Ogg, J.G. and Smith, A.G. (eds) (2004) *A Geologic Timescale 2004*. Cambridge University Press, Cambridge.

Hooker, J.J., Cook, E. and Benton, M.J. (2005) British Tertiary fossil mammal GCR sites. In Benton, M.J., Cook, E. and Hooker, J.J. *Mesozoic and Tertiary Fossil Mammals and Birds of Great Britain*, Geological Conservation Review Series, No. 32, Joint Nature Conservation Committee, Peterborough.

Howse, S.C.B. and Milner, A.R. (1995) The pterodactyloids from the Purbeck Limestone Formation of Dorset. *Bulletin of the Natural History Museum of London (Geology)*, **51**, (1), 73–88.

Hudleston, W.H. (1876) Excursion to Swindon and Faringdon. *Proceedings of the Geologists' Association*, **4**, 543–54.

Joffe, J. (1967) The 'Dwarf' Crocodiles of the Purbeck Formation, Dorset: A Reappraisal. *Palaeontology*, **10**, (4), 629–39.

Kenrick, P. (2010) Piecing the past together. *Evolve* **4**, 38–9.

Kenrick, P. and Davis, P. (2004) *Fossil Plants*, London: Natural History Museum.

Kowallis, B.J., Britt, B.B., Greenhalgh, B.W. and Sprinkel, D.A. (2007) New U-Pb zircon ages from an ash bed in the Brushy Basin Member of the Morrison Formation near Hanksville, Utah. *UGA Publications*, **36**, 75–80.

Lim, J.-D., Martin, L.D., Back, K.-S. (2001) The first discovery of a brachiosaurid from the Asian continent. *Naturwissenschaften* **88**, 82–4.

Martin, R.E. (1999) *Taphonomy: A Process Approach*, Cambridge: Cambridge University Press.

McBain, A. and Nelson, L. (2003) *The Bounding Stream: A History of Teffont in Wiltshire*, Teffont: Black Horse Books.

Milner, A.C. (2002) Theropod dinosaurs of the Purbeck Limestone Group, southern England. In Milner, A.R. and Batten, D.J. (eds) (2002) *Life and Environments in Purbeck Times*, Special Papers in Palaeontology No. 68, The Palaeontological Association, London.

Milner, A.R. (2004) The turtles of the Purbeck Limestone Group of Dorset, southern England. *Palaeontology* **47**, (6), 1441–67.

Milner, A.R. and Batten, D.J. (2002) Preface. In *Life and Environments in Purbeck Times*, Special Papers in Palaeontology No. 68, The Palaeontological Association, London.

Milner, A.R. and Batten, D.J. (eds) (2002) *Life and Environments in Purbeck Times*, Special Papers in Palaeontology No. 68, The Palaeontological Association, London.

Needham, J.E. (2007) A preliminary interpretation of Upper Jurassic silicified plant fossils from the Portland Stone Formation of Chicksgrove Quarry, Wiltshire. *Wiltshire Archaeological and Natural History Magazine*, **100**, 1–20.

Norman, D. B. and Barrett P. M. (2002) Ornithischian dinosaurs from the Lower Cretaceous (Berriasian) of England. In Milner, A.R. and Batten, D.J. (eds) *Life and Environments in Purbeck Times*, Special Papers in Palaeontology No 68, The Palaeontological Association, London.

Rawson, P.F. (1992) Jurassic (Introduction) in Cope, J.C.W., Ingham, J.K. and Rawson, P.F. (eds) *Atlas of Palaeogeography and Lithofacies*, Geological Society Memoir No. 13, The Geological Society, London.

Reid, C. (1903) *The Geology of the Country around Salisbury*, Memoirs of the Geological Survey, His Majesty's Stationery Office, London.

Rößler, R. (2002) Between Precious Inheritance and Immediate Experience-Paleobotanical Research From the Permian of Chemnitz, Germany. In Dernbach, U. and Tidwell, W.D. (eds) *Secrets of Petrified Plants: Fascination from Millions of Years*, D'ORO Publishers, Heppenheim.

Rothwell, G.W. and Holt, B. (1997) Fossils and Phenology in the Evolution of *Ginkgo biloba*. In Hori, T., Ridge, R.W., Tulecke, W., Del Tredici, P., Trémouillaux-Guiller, J. and Tobe, H. (eds), *Ginkgo biloba: A Global Treasure*, Springer.

Royer, D.L., Hickey, L.J. and Wing, S.L. (2003) Ecological conservatism in the 'living fossil' *Ginkgo*, *Paleobiology* **29** (1), 84–104.

References

Salisbury, S.W. (2002) Crocodilians from the Lower Cretaceous (Berriasian) Purbeck Limestone Group of Dorset, Southern England. In Milner, A.R. and Batten, D.J. (eds) *Life and Environments in Purbeck Times*, Special Papers in Palaeontology No 68, The Palaeontological Association, London.

Sellwood, B.W., Scott, J. and Lunn, G. (1986) Mesozoic basin evolution in Southern England, *Proceedings of the Geological Association*, **97** (3), 259–89.

Shindler, K. (2010) The first lady of fossils. *Daily Telegraph*, 15 June.

Stewart, W.N. and Rothwell, G.W. (1993) *Paleobotany and the Evolution of Plants* (2nd edition), Cambridge University Press.

Stockey, R.A. (2002) A Reinterpretation of the Cerro Cuadrado Fossil 'Seedlings', Argentina. In Dernbach, U. and Tidwell, W.D. (eds) *Secrets of Petrified Plants: Fascination from Millions of Years*, D'ORO Publishers, Heppenheim.

Taggart, R.E. and Cross, A.T. (1997) The Relationship between Land Plant Diversity and Productivity and Patterns of Dinosaur Herbivory. In *Dinofest International Proceedings*, 403–16.

Taylor, T.N., Taylor, E.L. and Krings, M. (2009) *Paleobotany: The Biology and Evolution of Fossil Plants*, Academic Press.

Thomas, B.A. (1991) The Study of Fossil Ferns. In Camus, J.M. (ed.) *The History of British Pteridology*, The British Pteridological Society.

Thomas, B.A. and Batten, D.J. (2001a) The Jurassic palaeobotany of Yorkshire. In Cleal, C.J., Thomas, B.A., Batten, D.J. and Collinson, M.E. *Mesozoic and Tertiary Palaeobotany of Great Britain*, Geological Conservation Review Series, No. 22, Joint Nature Conservation Committee, Peterborough.

Thomas, B.A. and Batten, D.J. (2001b) The Cretaceous palaeobotany of Great Britain. In Cleal, C.J., Thomas, B.A., Batten, D.J. and Collinson, M.E. (2001) *Mesozoic and Tertiary Palaeobotany of Great Britain*, Geological Conservation Review Series, No. 22, Joint Nature Conservation Committee, Peterborough.

Thomas, B.A. and Batten, D.J. (2001c) The Jurassic palaeobotany of Scotland. In Cleal, C.J., Thomas, B.A., Batten, D.J. and Collinson, M.E. (2001) *Mesozoic and Tertiary Palaeobotany of Great Britain*, Geological Conservation Review Series, No. 22, Joint Nature Conservation Committee, Peterborough.

Tidwell, W.D. (1990) Preliminary Report on the Megafossil Flora of the Upper Jurassic Morrison Formation. *Hunteria*, **2**, (8), 1–11.

Tidwell, W.D. (1998) *Common fossil plants of western North America* (2nd ed.). Washington DC: Smithsonian Institution Press.

Tidwell, W.D. and Medlyn, D.A. (1992) Short shoots from the Upper Jurassic Morrison Formation, Utah, Wyoming, and Colorado, USA. *Review of Palaeobotany and Palynology*, **71**, 219–38.

Vakhrameev, V.A. (1991) *Jurassic and Cretaceous floras and climates of the Earth*, Cambridge: Cambridge University Press.

Van Konijnenburg-van Cittert, J.H.A. (2008) The Jurassic fossil plant record of the UK area. *Proceedings of the Geologists' Association*, **119**, 59–72.

Weishampel, D.B., Barrett, P.M., Coria, R.A., Le Loeuff, J., Xing, X., Xijin, Z., Sahni, A., Gomani, E.M.P. and Noto, C.R. Dinosaur Distribution. In Weishampel, D.B., Dodson, P. and Osmólska, H. (eds) (2004) *The Dinosauria*, University of California Press.

Wimbledon, W.A. (1976) The Portland Beds (Upper Jurassic) of Wiltshire. *Wiltshire Archaeological and Natural History Magazine*, **71**, 3–11.

Winchester, S. (2001) *The Map that changed the World*, Penguin Group: Viking.

INDEX

Note: page references in *italic* type refer to illustrations.

abscission 132
acknowlegdements 27-8
Africa 51, 67
algae, algal 36, 49-50, 156, *156*, 173, 187
alligators 25
allochthonous 90
allosaurids 152
allosauroids 61, 116, *153*
Allosaurus 24, 25, 61, 116, 152, *153*
Alpine mountains 67
Alvin, K.L. 44
Amblotherium 153
Americas 51
ammonites *70*, *72*, 155
Ammonites giganteus 72
Amphibia 25
amphibians 24, 54, 56, 58, 64
Angiospermae *18*, 22
angiosperms 15, 17, 19, 22, 24, 40, 94, 112, 124-5, 197-8
Ankylosauria 62
ankylosaurs 24, *25*, 62, 152
Anning, Mary 68, 82
anticline 67
Anura 25
Apatosaurus 149, 151
Araucaria 191
  araucana 23, *38*
  mirabilis 39, 140
  sphaerocarpa 34
  sp. 139
Araucariaceae 21-2, 23, 34, 38, 198
araucarians *38*, 39, 48, 64, 104, 128, 140-1
Araucarioxylon 48, 128, 140

Araucarites 38, *191*
  sizerae 39, 99, 140
  sp. 94
Argentina 26, 39, 48, 52, 112, 136-7, *191*
armour 175-6, 179-81, 183, 189
Asia 21
Astin, T.R. 74-7, 81, 85, 146
Atlantic Ocean 51, 121, 142, 143, 198
Atoposauridae 57
atoposaurids *84*, 115
Australia 19, 21, *156*
  South 50
autochthonous 90, 187
Aves 25
Avon, River 67
Aylesbury 59

badgers 109
barite concretions 174
bark 76, 162, *163*, 165, 173, *176*, 185, *186*, 190
Barker, D. 49
Barremian 149
Basement Bed 70-1, *71*
Bass, Charles 125-6, 129
BBC 167
  Natural History Unit 167
Beckles, Samuel 63
'beef' 174
beetles 54, 187
Behunin, Homer and Joanne 123
Behuninia 126, 128, 131, 141, 146, 163, 169, *170*
  bassii 129, 132

- 211 -

joannei 124, 125-6, 129-30, *130*, 134-6, *136*, *191*
provoensis 129-35, *133*, *136*, *144*, 160, *191*
scottii 129, 132
sp. 133-4, 163
Belgium 82
Benett, Etheldred 68-9, *70*, 72, 77, 155
bennettites 19, 22, 31, 33, *35*, 36, 50, 53, 64, 97, 141-2, *141*, 175-6, 179, 182
trunks 31, 128
Bennettitales 18, *18*, 19, 33, 198
bentonitic ash 146
Bernissart 82
Bernissartia 57, 77, 81-2, *84*, 115
cf. *Bernissartia* 57, 84
Berriasian 30, *31*, 114, 117, *117*, 152, 155, *191*
binomen 129
bioturbation 88
birds 24, *25*, 152
bole 184-5
bones 55, 60-3, 73, *74*, 77, 80, 84-5, 87-9, 90, 92, 93, 114, *147*, 148-50
Brachiosauridae 149-50
brachiosaurids 60, 89, 116, 151
brachiosaurs 149, 151
Brachiosaurus 151
*altithorax*149
Brachyphyllum 37, 128, 140
sp. 37, 99
brackish 30, *55*, 56, 64, 119, 144
bracts 190
brash 79
Brigham Young University 123, 125-6
Bristow, Henry W. 55
Britain 54, 82, 96, 115, 155
British Geological Survey (BGS) 60, *71*, 78-9
British Isles 121
Brittany 121
Brontosaurus 149
Brown and Bugg 37-8, 99, 102-3, 108, 122-3, 139-40
Brown, Bugg and Costin, Misses 37
Brushy Basin Member 123, 131, *133*, 136, 139-40, *141*, 142-8
Buckinghamshire 58, *59*, 61, 89, 116, 149
Buckland, William 29, 31-2, 35, 61, 148, 175
buds 129, 132, 135, 163, 165, 169, *170*, 179-81

bugs 187
Bugle Pit *59*, 60-1
Building Stones, Lower, Main or Chief 69, 71, *71*
Building Stones, Upper 70-2, *71*
buttresses 185, *186*

caddis flies 54
Cainophytic 15, *15*
Cainozoic 14, *15*
calcification 92
calcite 87, 98, 107
calcium carbonate 92
Camarasauridae 149-50
camarasaurids 81, 149-51
Camarasaurus 149, 151, *191*
grandis 150, *150*
sp. 80, 150
camptosaurids 152
Camptosaurus 61, 116, 152
dispar 152
hoggii 152
carbonates 76, 99
Carboniferous 16, *51*, 99
carbonisation 91
Carnegie Museum of Natural History 147
Carnosauria 152
Carpolithes 37, 126
acinus 100, 111, *111*, 126, 138, *191*
cocos 105
gibbus 100, *191*
glans 101-3, *104*, *191*
rhabdotus 105-7, *106*
rubeola 101-2, *101*, 107-9, *108*, *191*
westi 37-9, *38*, 103, 139-40, *191*
Carpolithus 126, 128
Lindleyanus 38
provoensis 126, 129, 131
radiatus 126, *127*, 129
sp. 126
Carruthers, William 33
cataphylls 176-7
Caucasus 51
Caudata 25
Cayton Bay 17
Caytonia 112
Caytoniales 17, *18*, 197
caytonias 17, 112
cedars *18*, 21
cellulose acetate peeling 99

* Index *

Central America 19
Cephalotaxaceae 21
Cerro Cuadrado 39, 136-7, 140, 146, 190, *191*, 193
chalazal cap 106
Chalbury Camp 42-3, 171
chalcedony 52, 76
Chalky Series 69, 71-2, *71*, 79
Chandler, Marjorie E.J. 95, 122-6, *124*, 128-9, 131, 134-6, 138-40, 149
Channel 59
  Basin 120
charcoal 50, 109
Charophyte Chert 36, 38, 100, 104, 107, 139, *191*
charophytes 36, 50
Cheirolepidiaceae 22, 40, 45-6, 168, 192, *192*, 198
chelonians 77
chert 35-7, 75-6, 89, 92, 146, 156, 190, 197
Cherty Freshwater Member 53, 56-7, 60, 62, 65, 116, *117*, 122, 132
Chicksgrove 75, 77
  Member 71, *71*, 78
  Plant Bed 73, 75
Chilmark 26, 35, 68, 70, *71*
  Member *71*, 72
  Ravine 34-5, 68-9, 72, 118, 155-6
  Stone 68
China 12, 51-2
Chondrosteosaurus 149
choristoderes 114
chromosomes 96
chronostratigraphy 118
Circoporoxylon 48
circumscissile dehiscence 106
cladotheres 63, 117
Classopolis 36
  sp. 42
Classostrobus 42
  sp. 42
Clements, Roy 119
Cleveland Basin 33, 37, 97
climate
  arid 51, 58
  aridization 51, 145
  drier and seasonally erratic 50
  equable 145
  hot 196
  Mediterranean 48, 64, 113

more humid 54
Purbeck 32
seasonal, highly variable 145
seasonal sub-tropical 109
semi-arid 50-2, 143
semi-arid seasonally unstable 107
sub-tropical 64, 109
climbing plants 172
clones 136-7
clubmoss 16, *18*
fir 16
Coelurosauria 152
coelurosaurs 81
Colorado 26, 128, 131, 148
compression 20, 32, 67, 76, 88, 101, 137, 163, 167, *168*, 169, *170*, 172, 175, 179, 181, *181*, 184
compressions 33, 37, 41, 97, 128, 141, *170*, 171, 200
cones 17, 21, 22, 34, 38-40, *38*, 42, 48, 99, 104, 113, 126, 128, 139-40, 175-6, 179-80, 182, 184-5, 190-2, *191*, *192*, 193
cheirolepidiaceous 192
Coniferales *18*, 20, 106, 108, 126, 141
conifers 16, 21, 22, 26, 33, 36-7, 39-40, 42, 45, 48, 50, 53, 64, 80, 92, 97-8, *101*, 106, 112-3, 128, *130*, 140-1, 168, *170*, 172, *173*, 182-92, *186*, *188*, *191*, *196*, 198, *199*, 200
cheirolepidiacean 43
cheirolepidiaceous 44, 109, 140
cordaitalean 141
  seeds 80
Conifers, Age of *15*, *15*, 22
continental drift 121
coprolite 11, *188*
Coram, Robert 54
cortex 129
Corystospermales 17, *18*, 197
corystosperms 17, 112
Cranborne Chase 196
Cretaceous 14-15, *15*, 30, 63, 69, 82, *114*, 119, 140, 146, *147*, 155
  Early (=Lower) 82, 177
  Lower 15, *15*, 22, *31*, 37, 42, 44-5, 60, 75, 83, 89, 112, 115, 149, 197
  pre- 24
  Upper 15, *15*, 109, 115, 196
Crocodile Bed 115

crocodiles 24, 25, 54, 57, 64, 65, 68, 74, 81,
    84, 85, 87, 89, 115-6, *117*, 187, *191*,
    193, 198, 200
  dwarf 57, 64, 187, *188*, *191*
  'mesosuchian' 25
  Swanage 57
Crocodilia 25
crocodilians 73, 80, 148
  dwarf 77
crustaceans 53
Cryptodira 57
CT scanning 175, 192, *192*
ctenochasmatids 57-8, 115
Cupressaceae 21, 22, 34, 45-6, *47*
Cupressinocladus 34, 42, 46
  valdensis 42, 44-5
Cupressinoxylon 21
Cupressocyparis leylandii 34
Cupressus 40
  macrocarpa 45
cupules 102, 112
cuticles 42
Cuvier, Georges 82
Cycadales 18, *18*, 33, 125, 198
Cycadella 142
Cycadeoidales *18*, 19
Cycadeoidea 128, 141-2, 175, 177, 184
  gigantea 33, 177, 180
  megalophylla 31
  microphylla 31, 35
  wyomingensis 142
cycadeoids 33, 128, 142, 177, 179
Cycadeospermum schlubergi 106
Cycadites trunks 31
Cycadophyta 18, *18*, 19, 125
cycadophytes 18, 19, 24, 106, *106*, *111*, 114,
    124, 125, 134, 141-2, 149, *160*, 165,
    175-7, *178*, 179-85, *181*, *183*, 187-9,
    191, 192, *196*, 197-8, *199*, 200
  age of the 19
  trunks 19, *20*, 49, 141-2, *141*, 164, 175-8,
    178, 181-2, 198
cycads 18, 20, 33-4, 142, 174, 179
  'true' 18, *18*, 19, 33, 125, 175-6
Cycads, Age of 19
cypresses 40-1, 45-6
  leyland 47
Czech Republic 109
Czekanowskiales 17, *18*, 197

Damon, Robert 49
Darwin, Charles 32, 195
Davis, Paul 24
Dayvault, Richard D. 131-2, 135-7, 139
defossilization 96
Deleyvoryas, T. 142
dendrites 174
diagenesis 75
Dinosaur National Monument 26, 147-9
dinosaurs *Passim*
  armoured 24, 62, 77, 79, 81, 152
  bird 12
  bird-hipped 24, 25, 61, 152
  eggshell 116
  footprints 55, 58, 60, 62-3
  graveyards 80, 147
  lizard-hipped 24, 58
Dinton 197
Diplodocidae 151
diplodocids 81, 151
Diplodocoidea 151
Diplodocus 15, 24, 25, 151
  sp. 80
dirt beds 31, 34, 36, 41, *55*, 62, 79, 156
docodonts 63, 153
Dorchester 32, 47
Dorset *Passim*
  County Museum 32, 47
Dorsetisaurus 57, 148
Dorsetochelys 56, 147
Douglass, Earl 148
dragonflies 54
dromaeosaurids 60-1, 116, 152
dromaeosaurs 62, 152
dryolestids 117, 153
Durlston Bay 54, 57, 62-3, 65, 84, 117, 132,
    187
Durlston Formation 31, 54, 57, 115
Dykes, Trevor 83

Eastern Pyrenees 44, *158*, 167, *168*
Echinodon 152
  becklesii 62
  cf. *Echinodon* 152
  sp. 80
endosperm 110
England 17, *38*, 44, 69, 77, 84, 97, 142, 147,
    196, 198
  eastern 121
  south-central 26, 29, *70*, 80, 100, 120, 122,

* Index *

143, 153, 197
south-east 84, 96, 120
southern 22, 45, 49, 51, 59-60, 67-9, 79, 81-2, 90, 97-8, 115, 119, 121, 134, 137, 142-3, 147-9, 192
Ensom, Paul 56, 58, 63, 116
Eocene 94-6
epiphyte 164
Equisetales 18, 37
Equisetopsida 18
Equisetum 197
  mobergii 37, 99
Eupantotheria 82, 117
Eurasia 51
Europe 26, 60, 67, 73, 98, 122, 140, 149, 155, 196
  central 52
  northern 59
evaporates 51
evaporite 51
  pseudomorphs 50
exoskeleton 186

fabrosaurids 81
facies 118-9
Feather Quarry 60
ferns 17, 18, 22, 36, 50, 64, 128, 131, 140, 175, 197, 200
  pinnules 53
Filicopsida 18
fire 50, 103, 107, 109, 113, 165
fish 55-6, *55*, 64, 72, 85, 93, 155
  bony 73
  Devonian 193
Fitton, William H. 32, *32*, 46-7
'flame' effect 180-1, *181*
flies 54
flint 36
Flowers, Age of 15, *15*, 197
Fonthill Bishop 26
Fonthill Gifford 26
Form A, B, C 105
form-genus 34
Fossil Forest (Purbeck Dorset) 30, *30*, *32*, 32-4, 36, 39, 40, 45-50, *51*, 52, 54-6, 58, 90, 96-8, 107, 112, 122, 131, 139-42, 145, 155, 171, *173*, 175, 190, *191*
Fossil Forest (Purbeck Wiltshire) 20, *141*, 155-93, *156*, *158*, *173*, *181*, *183*, *186*, *188*, *189*, *191*, *192*, 196, *198-9*, *199*

France 51, *106*, 120, 145, 150
Francis, Jane E. 39, 40, 42-3, 45-50, 52, 75, 113, 145
frankincense 96
frass 170, *173*
Fremont site 123, *124*, 128, 149
freshwater 30, 50, *55*, 56, 58, 60, 64, 115, 119, 143-4
frogs 24, 25, 56, 64, 114, 147, *147*
fronds 175, 178, 198
fructifications 125-6
Fruitadens 152
fruits 17, 94-5, *95*, *108*, 110
fungi 171

gastrolith 110
gastropod micrite 59, 75-6, 86-9, 115-6, 118, *120*, 151, *153*
gastropods 75, 80, 85, 87, 119, *188*
Gault Clay 197
Geological Conservation Review (GCR) 97-9, 114, 116, 119
  site 79, 82-3, 98
Geological Society of London Library 69
germination 103, 106-7, 109, 110, 113
giant kauri tree 21
Ginkgo 108-9
Ginkgo biloba 20, 108, 128, 136
Ginkgoales 18, 20, 108, 198
ginkgophytes 108-9, 128, 200
ginkgos 24, *108*
gizzard 110
glauconite 72
Glyptops 56, 147
Gnathosaurus 57, 81, 115
Gnetales 18, 19
God Nore 41
Goniopholidae 57
goniopholids 81
Goniopholis 57, 77, 115, *117*, 148, *191*
grape 96
grasses 22
Great Dirt Bed 30, 34, 36, 41-2, 50, 75-6, 156
Guimarota 73, 83
Gymnospermophyta 17, *18*, 128
gymnosperms 17, *18*, 20, 24, 125

Haddenham Formation 59, 61
halophytes 109

- 215 -

Hampshire 68
Harris, Tom M. 37, 124
Hastings, Barbara Rawdon, Marchioness of 68, 193
Hatch, H. Steven 131-2, 135-7, 139
Hayes, Peta 165
Hebrides 121
Helochelydra 56, 114-5
Henry Mountains *124*, *125*, *127*, *133*, 139
herbivores 24, 62, *108*, 110, 134, 184
heterodontosaurids 62, 80
heterodontosaurs 152
hilum, hilar 100-2, 105, 107, 110
Hillistrobus axelrodi 126, 140
holotype 100-3, 105
horsetails 17, *18*, 37, 39, 49, 50, *51*, 64, 99, 140, 197, 200
Hudleston, W. H. 56
Hutton, James 12
hypersaline 30, 49, 52-3, 144, 165, 168, 171

Iberian Peninsula 115, 143
Ichthyosauria 25
ichthyosaurs 25, *25*, 62
Iguanodon 11, 15, 24, 81-2, 116
  hoggii 61, 152
iguanodonts 25, 116
impressions 33, *47*, 49, 98
India 96
Inferior Oolite 38
insects 54, 64, 155, 171, *173*, 186-7, *188*, 198
  boring 19, *173*
in situ 30, 49, 53, 63, 90, 154, *166*, 171-2, 176, 184, 187, 189-90, *191*
integument 100-2, 108
International Commission on Stratigraphy 14
intertidal 30, 64
invertebrates 53-4, 58, 60, 64, 119, 121, 171

Japan 18
Jensen, James A. 123
Jensenispermum 110, *111*, 126
  redmondi 125-6, 138-9
JNCC 83
joint firs *18*, 19
junipers *18*, 21, 45-6
Juniperus oxycedrus 45
Jurassic *Passim*

Late (=Upper) 45, 98, 143
Lower 15
Middle 15, 21, 33, 37, 50, 61, 97-8, 109, 112, 141, 200
pre- 29
Upper 14, *15*, *31*, 33, 37, 44, *44*, 50, 67, 69, 70, 79, 82, 96-8, 105, 109, 113-5, 122, 129, 137, 142, 145, 154, 165, 187, 198
Jurassic Coast 26, 47, 64, 112
Jurassic Park 61, 96

Karst topography 188
Kenrick, Paul 24, 164-5, 167-8, 170, 172
Kent 94, *95*
Ketchum, Hillary 167
Kimmeridge Clay 69
Kimmeridgian 67, 73, 143, 149
knots 159-60, 164-5, 169
Ktalenia 112

lacewings 54
lagoon, lagoonal 30, 36, 50, 52-4, 59, 64, 73, 75, 77, 121, 155, *156*, 165, 168, 171, 197-8, *199*, 200
  hypersaline 168, 171
Lake T'oo'dichi' 143
Landers Quarry 53
larch 130
Larix 130
larvae 170
  beetle 19, *173*
laser level 182
laser scanning 168
leaves 17, 19, 33-4, 41-2, *47*, 97, 99, 128-9, 141, 175-6, 184, 197-8, 200
lepidosaurians 77
lepidosaurs 54, 57, 80
Lepidotes 85
Leyland cypress 47
lignite 75, 171
lignotubers 135-8, *138*, 163-4, *163*, *189*, 190, *191*
limestone platform 188
lithostratigraphy 118
lizards 24, *25*, 57, 61, 64, 80, 114, 148, 200
London 32, 35, 61, 89, 100, 125, 139, 167
London Clay 94, 96
Lourinhasaurus 149
Low Countries 121
Lower Chicksgrove 26, 67

Lower Dirt Bed 30, 48, 171
Lower Greensand 22, 197
Lulworth 33, 48, 53, 62
Lulworth Formation 30, *31*, 54, 56-8, 62, 115, 122
Lycophyta 16, *18*
lycopods 50, 64, 200
Lycopsida *18*
Lyme Regis 68

macrofossils 36, 50, 53
Macronaria 149, 151
magnolia 96
maidenhair tree *18*, 20, 108-9, 128
Malaysia 40
mammal bed 73
Mammal Bed 63, 73, 115
Mammalia 25
Mammaliaformes 25, 153
mammals 25, *25*, 54, 63-4, 72-4, *74*, 77, 80, 82-3, 110, 116-7, 152-3, 155, *191*, 200
Mammals, Age of 14, *15*
mangrove 96
Maniraptora 25
maniraptors 24
Mantell, Dr. Gideon 11, 14, 20, 24, 32, 60, 82, 84, 148
Mantellia (Cycadeoidea) microphylla 35
Marly Freshwater Member 53, 62-3, 122
Medlyn, David A. 128, 130-2, 134-6
Megalosaurus 15, 61
sp. 80
megaspores 80
Mere fault 120
mesic 50
Mesophytic 15, *15*, 22, 23
Mesozoic 14-17, *15*, *18*, 19-20, 22, 94, 97, 108-9, 124, 140, 153, 165, 179, 192
seed ferns 17, *111*, 112
micropyle, micropylar 100-2, 105-7, *101*, 110-2
microspores 80
Middle Asia 51
mineralisation 110, 173
miospores 36, 50, 140, 200
Moab 135
molluscs 53, 69, *70*, 85
Mongolia 51
monkey-puzzle tree *18*, 21, *23*, *38*, 48
Montana 128, 140

mormon teas *18*, 19
morphogenus 34, 36-7, 40, *41*, 42, 46, 48, 99, 112, 126, 128, 140, 146
morphotaxa 90, 109, 134, 136-7, *138*
Morrison Formation 74, 83, 123, 125-6, 131, 136-7, 139-40, 142-53, *150*, *153*, 163, 174, 177, 179-80, 190, *191*
mosasaurs 25
Mount Ellen 125-6, 129, 131-2, 139
mountain pine *44*, *98*, *158*, *167*, *168*
mudflats 30, 58, 63-4, 143, *199*
Multituberculata 82, 116
multituberculates 63, 73, 83, 110, 117, 153
mycorrhiza 171

Nadder, River 26, *51*, 67, 195
valley 35
Naish, Darren 149
Natural History Museum (NHM) 24, 26, 61, 89, 100, 110, 112, 114, 116, 125, 132, 139, 151, 164-8, 172, *192*, 193
Nature Conservancy Council 80
Neogene *15*
Newfoundland 143
New Zealand 21
newts 25
Nodosauridae 80
nodosaurs 24, 25, 62, 81, 116, 152
nomenclature 45
non-diplodocoid non-titanosauriform eusauropods *150*, 151
Norris, G. 36
North America 74, 115, 131, 179
North Dakota 108
Northern Hemisphere 50, 108
Norway spruce 43
nucellus 104
Nuthetes 60, 77, 116, 152
cf. *Nuthetes* 116, 152
destructor 60, 62

Oakley Marl Member *71*, 79
Oakley Quarry 55
Obligaster fittoni 54
opal 52, 76
operculum 106
Opisthias 148
Ornithischia 24, *25*, 61, 80, 152
ornithischians 62, 80
Ornithocheirus 58

ornithopods 77
orthotropous 100-1
ostracods 60
ovules 103-4, 112
Owen, Richard 60-3, 148
Oxford 29, 73
Oxford University Mus. Nat. Hist. 62
Oxfordian 105
Oxfordshire 61

palaeocarpology 95, *95*, 113
Palaeocene 108
Palaeogene *15*, 94
Palaeophytic 16, 99
palaeosol 30, 56, 63, 75, 156-8, 160, 171-3, 176-7, 179, 182, *183*, 184-8, *188*, 190, *192*, *196*, *199*
Palaeozoic 16
palm 20, 96
Pangea 51
pantotheres 73, 83
Pararaucaria 190, 192
Pararaucariaceae 22
parasitic relationship 174
Paris Basin 120
Patagonia 137, 142, 146, 190, 192
Pelorosaurus 60-1, 89, 116, 149
  cf. *Pelorosaurus* 59, 89, 149, 151
  sp. 60
Pennines 17
periderm 172
permineralisation 91, 98-9
petioles 175, *178*, 179
petrifactions 26, 31, 33, 91, *92*, 94, 97-9, *124*, 127
petrifications 91, 187
phloem 175
physical dormancy 103
Picea abies 43
Pinaceae 21, 22, *130*, 198
Pinales *18*, 20
pines *18*, 21, 46, 128
  needles 131
pinnae 19
Pinopsida *18*
Pinus mugo 44, *98*, 167
pith 129, 132, 135, 175, 180-1, *181*, 184, 189
Plagiaulax 63
plankton 196
plant and reptile bed *Passim*

plants *Passim*
  bed 73
  flowering 15-16, *18*, 22, 24
  kingdom 16, *18*, 22
  vascular 16, *18*
Plataleorhynchus 57
Plesiosauria 25
plesiosaurs 25, *25*
Pleurosternon 56
plum-yew, Chinese 21
Podocarpaceae 21, 48
podocarps 21, 48, 64
point bar deposits 93
pollen 19, 36, 40, 42, 100, 200
pollination 111
pollinators 19
Poole 46
Portesham Quarry 36, *38*, 39, 48-50, *51*, 79, 96, 99, 100-6, *101*, *104*, 113, 118-9, 122-3, 139-40
Portland, Isle of 29, 31-6, *32*, 41-3, 48, 79, 97, 156, 171, 177, 179-80
Portland Beds 71, 77, 89, 119
  Lower 69, 71, *71*
  Upper *31*, 69, *71*, 78
Portland Sand 69, *71*
Portland Sand Formation 71, *71*, 78
Portland Stone 26, 30, *31*, 55, 67, 69, *71*, 72, 77-9, 81, 120
Portland Stone Formation 27, 29-30, *31*, 55, 59, *59*, 62, *70*, *71*, 72, 75, 78, 90, 92, 113, 118, 155
Portlandian 60, 77, 98, 150
Portsmouth University 12
Portugal 73, 83, 143, 149
preventitious buds 165
Protocupressinoxylon 40, 45-6
  purbeckensis 40, 42, 45, *47*, 48, 145, 157, 192
Pteridophyta 17, *18*
Pteridospermophyta 17, *18*
pteridospermophytes 24, *111*, 197, 200
pterodactyloids 148
Pterodactylus 81-2
Pterosauria 24, *25*
pterosaurs 54, 57-8, 64, 73, 74, 77, 80-1, 89, 90, 115, 148, 200-1
pupa 186-7
pupation chamber 171
Purbeck Beds 29, *31*, 36, 56, 119

Index

Lower 30, *31*, 42, 58, 62, 69, *71*
Middle *31*, 57, 69
Upper *31*, 57, 69
Purbeck Formation 29, *71*, 79
Purbeck Group 29, 60
Purbeck, Isle of 29, 55-6, 60
Purbeck Limestone 26
Purbeck Limestone Formation 29, 79
Purbeck Limestone Group 29, 30, *31*, 33-4, 35, *38*, *47*, 53-4, *55*, 56-8, *59*, 60-2, 65, 75, 79-80, 82-3, *84*, 85, 89, 93, 114-19, *117*, 121, 132, 145-9, *147*, 152-5, 186-7, 197-8, 200
PVA glue 86, *87*, *136*, 164
pyrite 92, 94
pyritisation 92
Pyt House 68, 77

quartz 52, 76
quartzine 52
Quaternary 15
quillworts 16, *18*, 197

rachis 125
radioactive dating 146
Ragstones 69, *71*, 72, 79
ramentum, ramenta 175, 177, *178*, 179-80, *181*, 182, 184, 189
raptors 61
Reading University 37
Reid, Eleanor M. 95, 123
redwoods *18*, 21
Regent's Park 32
replacement 91
reptiles 24, 54, 56, 58, 61, 64, 72-3, *74*, 77, 80-1, 83, 87-9, 92, 110, 114, 116, 155, 187
  eggshell 116, 187, *188*
  flying 24, 25, 57, 89, *90*
  marine 54, 62, 68
Reptiles, Age of 14, *15*, 16, 82
Reptilia 25
resin canals 172
rhamphorhyncoids 58, 148
rhizomes 175, 185
Ridge 35
rings 40, *41*, 48, 113, 145, 184-5
  annual 112
  false 112
Rocky Mountains 143

rootlets 171-2, 174, *176*, 185-6
roots 31-2, *41*, 42-3, 53, *138*, 166-7, 171-5, *176*, 178, 184-6, *186*, 196
Royal Botanic Society 32
Ruflorinia 112

sago palm 18, *18*
salamanders 24, 25, 56, 64
salamandroids 147
Salisbury 67, 73, 78
  Cathedral 68
  library 85
  Plain 196
de Saporta, G. 105
saprophytes 174
sarcotesta 108-110
saurian 77
Saurischia 24, 25
saurischians 58, 60, 80
Sauropoda 80
sauropods 24, 25, 58, *59*, 60, 77, 80-1, 88-9, 116, 147, 149-51, *150*, *191*, *199*, 200-1
sclerotesta 108, 110
Scotland 97, 121
  north-east 97
Secondary 14, *15*
secondary xylem 129, 132, 134-5, *158*, 172
seedlings 136-8, *138*, 190
seeds 17, 21, 26, 34, 37-9, *38*, 50, 64, 85, 90, 93-6, *95*, 99-100, *101*, 102-113, *104*, *106*, *108*, 122, *124*, 125-6, *127*, 128-9, 131, 134, 139-40, 149, 190
  capsules 17, 111-112, *111*, 126, 138
  chamber 111-12
  enclosed 22
Sequoia sp. 126, 140
sequoias 126
Seward, Albert C. 33, 38, 177, 180
Shaftesbury 197
sharks 73
shellfish 57
Sheppey, Isle of 94-6, *95*, 123
shoots 41, 42, *47*, *81*, *87*, 90, 94, 99, 131-2, *133*, 134-8, *138*, 140, 144, 160, *163*, 164, 175, 180, *191*
  conifer 34, 50, 53, *78*
  long 128, *130*, 131-4
  short 21, *127*, 128-9, *130*, 131-6, *136*, *138*, 139-41, *144*, 145-6, 149, 154, 160, 163, *163*, 169, *170*, 189, 190

- 219 -

spur 17, 129
shrubs 17
Sigogneau-Russell, D. 63
silica 36, 52, 76, 78, 92, 92, 146, 167, 171, 180, 186
silicic acid 52
silicification 31, 37, 40, 52-3, 76, 91, 92, 100, 110, 142, 146, 157-8, 160, 167, 177-9, 184
Skye, Isle of 97
Smith, William 69
snails 187, 198
snakes 24, 25
Solemydidae 114, *114*, 191
solemydids 114-5
South Africa 19, 52
South America 22, 179
South Dakota 128
Southampton 73
Southern Hemisphere 23, 50, 109
Spalacotherium 63
Sphenodontia 25, 54
sphenodonts 25, 148
Sphenophyta 16-17, *18*
sponges 69
  spicules 52, 197
spores 36, 75
Squamata 25, 54, 57
squamates 148
SSSIs 60, 73, 77, 80, 83, 85, 114
Stegosauria 79
stegosaurid 152
stegosaurs 24, 25, 81, 152
Stegosaurus 143, 152
Steiner, Richard 129
Steinerocaulis 128, 141, 146, *189*, 190
  radiatus 127, 129-30, 135, 137, *191*
Stonesfield 61
Stopes, Marie 22
Stratiotes 93
stromatolites 156-7, *156*, 162, 164-5, 187-8, 192, 198, *199*
stromatolitic limestone 49, 52, 157
strophiole 106
Sunnydown Farm 56, 58, 63, 114, 148
Supersaurus 151
Sussex 197
Sweden 37
Sweetman, Dr. Steve 12
Swindon 29, 56, 63-5, 77, 117, 119, *147*, 155

Switzerland 51
symbiotic relationship 174
Synapsida 25

Tanzania 83
taphonomy 91-2
Taxaceae 21, 198
Taxales 21
Taxodiaceae 21, 22, 46, 126, 128, 141, 198
Taxodineae 126
taxodioids, transitional 22, 46
tea 96
technotheres 63
tectonic movements 118, 120
teeth 26, 55, 57-8, 59, 60-3, *62*, *74*, 77, 80-2, *84*, *85*, 87-9, 93, 115-6, *117*, 147-53, *150*, *153*, 187, *188*, 191
  shark 84
Teffont Evias 26, 54, 56-7, 155
Tendaguru 83
tendrils 174
Tennyson, Alfred 195
Tertiary 14, *15*, 21, 68, 68, 84, 94, 95, 108, 124
Testudines 25
Texas 115
Thame 59
theriosuchians 81
Theriosuchus 57, 84, 115, 187
  cf. 84, *188*
Theropoda 80
theropods 24, 25, 60-1, 62, 64, 77, 81, 87, 89, 116, 152, 201
thin sectioning 179
thorns 134, *144*
Thyreophora 152
thyreophorans 152
Tidwell, William D. 126-8, 130-2, 134-6, 141
Tisbury 26, 27, 55, 67-9, 71, 78, 120, 195, 200
Tisbury Coral 69
Tisbury Member 71-2, *71*, 75, 79, 86, 89, 90, 92, 118, *120*, 134, 146
Tisbury Row 68
Titanites giganteus 72
Titanosauriformes 149
Tithonian *Passim*
toads 25
tortoises 25

Index

Town Gardens Quarry 63
trace fossils 171
trees 26, 39, 42, 44-5, 47, *47*, 53, 64, 113, 134, 136-7, 141, 145-6, 154, 157-60, *158*, *160*, 162-71, *163*, *166*, *170*, 174-5, 182-3, *186*, 192-3, 198, *199*, 200
   broad-leafed 22
   Christmas 43
   coniferous 31
   monopodial 43, *44*, 46-7
   stump 31-2, 35, *35*, 42-3, 46-7, 49, 52-3, 63, 112, 154, 157, 165-7, *168*, 169, 174-6, *178*, 179-80, 182, 187
   trunk 31-2, *32*, 39, 42-3, 46-7, 49, 53, 63, 112, 137, 154, 157-9, *160*, 162, 165, *166*, 167-9, 188, 192, *196*
Triassic 14, *15*, 112
Triconodonta 82, 117
triconodonts 63, 117, 153
troodontids 152
trophophylls 175
tuatara 25
turtles 24, *25*, 54, 56-7, 64, 65, 77, 89, 114-5, *114*, 116, 147, *191*
Tyrannosaurus rex 15

UK 17, 83, 86, 94, 97, 116, 131, 149, 152
Ukraine 51
Umkomasia resinosa 112
Upper Chicksgrove 26, 67-9, 72-3, 79, 82-3
Upper Chicksgrove Quarry 83, 85, 118, 150
Upper Greensand 197
USA 83, 127, 139, 143, 174, *191*
   southern 52
   western 26, 134, 140, *141*, 142-3, 145, 147-8, 153, 177, 192
Utah 26, 122-6, *124*, 127, 128, *130*, 131, 133-5, *133*, 139-40, 142-3, 145-6, 148-50, *150*, 153, *153*
Utah State University 128

Vakhrameev, V.A. 50, 145
Vale of Wardour 26, *27*, 29, 34, 56, *59*, 64-5, 67, 69, 71-2, 77, 84-5, 118, 155, 158, 196-7
velociraptors 62
vertebrates 54-5, 60, 63, 73, 77, 86, 90, 121-2, 147-8, 153, 155, 186

higher *25*
volcanic ash 52, 144, 146

Wales 69, 121
walnut 96, 101
Ward, Lester F. 19, 142
Wardour 26
Wardour Formation *71*, 78
Wardour Member 71, *71*, 78
wasps 54, 187
Watson, J. 44
Weald Basin 120, 197
Wealden 42, 44-5, 60, 75, 115, 197, 200
   Clay 198
Wessex Basin 120, 197
Wessex Formation 149, 197
West, Ian 36, 103
Weymouth 36
   Relief Road 53-4
whiptails 151
Wight, Isle of 12, 40, 93, 197
Wildlife and Countryside Act 73
Willow Springs 123
Wilton 67
Wiltshire *Passim*
   County Council 73, 80
   Inter-library lending 99
Wimbledon, W.A. 71-5, *71*, 77-9, 85
Wincanton 78-9
Wockley Member *70*, 71, 72-3, 75, 79, 118
Wollemi pine 21, *23*
wood 21, 32, 35-6, 39-42, *41*, 45-6, 48, 52, 76, 80, *81*, 88, 92, 94, 98, 112-3, 120, 20, 127-8, 132-4, *133*, 140, 145-6, 155, 157, 160, 163-4, 167, *168*, 169, *170*, 171, *173*, 174, 183-5, 188-90
   boring 170-1, *173*
   silicified 50
Woodward, H.B. 156
Wyoming 128-130, 140, 150

Xenoxylon 128
xeric 112
xylem 175

Yellow Cat 135
yews *18*, 21
Yorkshire 17, 21, 33, 37, 50, 97, 109, 112, 140-1, 200

www.ingramcontent.com/pod-product-compliance
Lightning Source LLC
Chambersburg PA
CBHW031142160426
43193CB00008B/218